THE SCIENCE OF
SERIAL
KILLERS

THE SCIENCE OF
SERIAL KILLERS

THE **TRUTH** BEHIND
TED BUNDY, LIZZIE BORDEN,
JACK THE RIPPER, AND
OTHER NOTORIOUS MURDERERS
OF CINEMATIC LEGEND

MEG HAFDAHL & KELLY FLORENCE
AUTHORS OF *THE SCIENCE OF STEPHEN KING*

Skyhorse Publishing

Skyhorse Publishing books may be purchased in bulk at special discounts for sales promotion, corporate gifts, fund-raising, or educational purposes. Special editions can also be created to specifications. For details, contact the Special Sales Department, Skyhorse Publishing, 307 West 36th Street, 11th Floor, New York, NY 10018 or info@skyhorsepublishing.com.

Skyhorse® and Skyhorse Publishing® are registered trademarks of Skyhorse Publishing, Inc.®, a Delaware corporation.

Visit our website at www.skyhorsepublishing.com.

10 9 8 7 6 5 4 3 2 1

Library of Congress Cataloging-in-Publication Data is available on file.

Cover design by David Ter-Avanesyan
Cover photos: DNA by Shuttershock; hand by Getty Images

Print ISBN: 978-1-5107-6414-9
Ebook ISBN: 978-1-5107-6415-6

Printed in the United States of America

We dedicate this book to the victims, known and unknown,
and to their families and friends.

CONTENTS

INTRODUCTION

Killers exist in every aspect of our collective experience. Fictional ones loom in the films we watch and the books we read. Real ones strike in our own hometowns. There is a profound moment in our lives when we come to realize that the Freddy Krueger of our nightmares pales in vicious comparison to the serial killers in the newspapers. Within these pages we explore that fractious line that wavers between fiction and reality. Hollywood and true crime have long been bedfellows, and through the lens of science we can come to further understand the murderers who have darkened history and the films they have inspired.

This book is filled with dark deeds; many real, and others pure fiction. Tread cautiously, as this hazy border may trip you up.

As always, our love for horror in all its iterations lies within. We *don't* love the true killers, but we sure do love when brave victims, families, and members of law enforcement are honored. Often, film gives way to better stories in which the final girl slays the monster, and for that we are grateful. Let the monsters be forgotten as we rise.

SECTION ONE
CRIMES OF THE PAST

CHAPTER ONE
The Legend of Lizzie Borden (Lizzie Borden)

I (Meg) grew up in the eighties and nineties, overstuffed with a diet of iconic horror villains. Freddy, Jason, and even the menacing eyebrows of Vincent Price led me down the familiar landscape of twentieth century horror. I was only eight, snuggled in a hotel bed with my mother, when I discovered that that women, too, had the power and nuance to be the foreboding presence in a shadowed room. While we watched a rerun of *The Legend of Lizzie Borden* (1975), a TV movie starring Elizabeth Montgomery from *Bewitched* (1964–1972), my mom divulged that the events unfolding were actually based on fact! The hazy images of Freddy Krueger and his vicious film brothers left my mind. The tropes I had already come to know, of women screaming in terror as they fled a maniac, had fallen at my feet. I was exalted by the thought that a woman, a real woman, in her fancy dress and with her flowery words, could succumb to the murderous, black heart within.

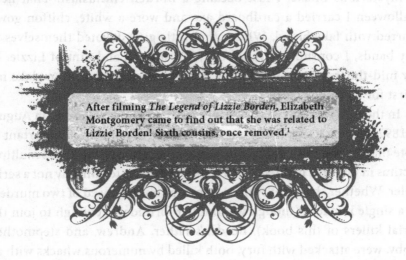

After filming *The Legend of Lizzie Borden*, Elizabeth Montgomery came to find out that she was related to Lizzie Borden! Sixth cousins, once removed.[1]

While today *The Legend of Lizzie Borden* may not be the most famous of Borden biopics, it was popular upon its release. Montgomery was nominated for an Emmy, and writer William Bast received the Edgar Award for a television film script. For many in the 1970s, it had been their first introduction to the history and evidence of the trial, rather than the sing-song rhyme made so popular:

> Lizzie Borden took an axe,
> And gave her mother forty whacks;
> When she saw what she had done,
> She gave her father forty-one!

Since the TV movie, there have been quite a few adaptations of Lizzie and her infamous axe. These include the Lifetime TV film starring Christina Ricci, *Lizzie Borden Took an Axe* (2014), as well as *Lizzie* (2018), in which Lizzie (Chloe Sevigny) and her family maid Bridget Sullivan (Kristen Stewart) have a steamy love affair. This lesbian plot point also appears in *Lizzie: The Musical* (2009). While there is no evidence to prove that Lizzie and Bridget were lovers, Lizzie Borden was a known lesbian in her small community of Fall River, Massachusetts. Thus, the possibility of a relationship has been a suspected motive for many.

After watching Elizabeth Montgomery inhabit the curious skin of mysterious Lizzie, I fast became a Borden enthusiast. That next Halloween I carried a cardboard axe and wore a white, chiffon gown marred with fake blood. While other little girls devoted themselves to boy bands, I consumed every book, TV special, and hint of Lizzie. In my mid-thirties I am still finding new books and theories to slake my thirst for all things Lizzie.

In the film versions of what occurred in the Borden house on August 4, 1892, there is a surprising amount of accuracy. It is first important to note that, by the FBI's definition, serial killers must have killed multiple victims in multiple incidents. Therefore, Borden is technically not a serial killer. Whether she is guilty or not, she was only accused of two murders in a single incident (though we find her formidable enough to join the serial killers of this book). Borden's father, Andrew, and stepmother, Abby, were attacked with fury, both killed by numerous whacks with an

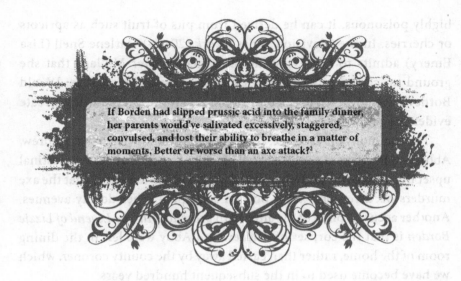

If Borden had slipped prussic acid into the family dinner, her parents would've salivated excessively, staggered, convulsed, and lost their ability to breathe in a matter of moments. Better or worse than an axe attack?[2]

axe. In *The Legend of Lizzie Borden*, Andrew Borden's (Fritz Weaver's) body is splayed out on a divan in exact reference to the infamous crime scene photograph (though his head is far less mutilated for the benefit of those tuning in to the TV movie). Other commonly known facts are also shared in the film, including the evidence given by a local druggist, Eli Bence (Olan Soule), who told the judge and jurors that "the day before the murders, Lizzie had come in shopping for prussic acid—a deadly poison."[3]

Poison has long been considered a woman's tool for murder. There are many known female serial killers who relied on this silent weapon, like Nannie Doss who poisoned eleven people, and Nurse Jane Toppan (page 62) who killed her own patients with lethal mixes of medicine. According to the *Washington Post*, women are seven times more likely than men to kill with poison. In the centuries before forensic scientists developed a way to detect nearly all substances in a victim's body, women used poison to annihilate their husbands, children, and bothersome neighbors. The *Post* article further extrapolates the data from the FBI: "Killers over thirty are more than twice as likely as younger killers to resort to poison."[4]

The fact that Borden sought out prussic acid the day before the murders has long been used as evidence of her guilt. Prussic acid, also known as hydrogen cyanide, is a colorless and flammable liquid that is highly poisonous. It can be derived from pits of fruit such as apricots or cherries. In the television series *Ozark* (2017–) Darlene Snell (Lisa Emery) admits to her dying husband Jacob (Peter Mullan) that she ground up cherry pits to create cyanide to poison his coffee. Would Borden have bothered to ask for prussic acid if she knew it would create evidence against her?

The testimony of Eli Bence, coupled with the known fact that Andrew, Abby, and Borden's sister Emma fell ill with a mysterious gastrointestinal upset the week prior to the axe murders, led many to believe that the axe murders occurred only after Borden attempted other deadly avenues. Another accurate piece of the aftermath presented in *The Legend of Lizzie Borden* is that the corpses of Andrew and Abby were left in the dining room of the home, rather than carted away by the county coroner, which we have become used to in the subsequent hundred years.

It made us wonder why this was a popular practice at the time.

The National Museum of Funeral History in Houston, Texas, is dedicated to exhibiting world practices of mourning the dead. Their exhibits include such varying topics as "The History of Cremation" to "Fantasy Coffins from Ghana." One of the museum's rooms recreates a Victorian parlor not unlike the Borden's living space. It depicts a mannequin in black veil, attending the dead in their own home. The museum's proprietors explain this culture of mourning in the nineteenth century on their website:

During the 1800s, determining that a person was actually dead was not as simple as it is today, as they didn't have the medical technology we do now to determine true death. During the days following a person's death, the body was closely observed for three days to make sure the person didn't wake from a deep sleep or illness before the funeral and burial—thus the term "wake" we use today for visiting/viewing the recently deceased. During the early twentieth century, funeral service practitioners transitioned from providing in-home services to establishing funeral homes, where bodies were transported and prepared for funeral services. It was during this time that parlors became known as "living rooms," because they were no longer used to display the dead.[5]

It's darkly humorous to believe that anyone might have questioned whether Andrew or Abby Borden were deceased after numerous blows with an axe. While it must have been obvious that they were dead, the traditions of the era remained. They were left downstairs in the heat of the summer beneath nothing more than thin sheets for days. In his piece for the *Chicago Tribune*, reporter William Hageman explains further:

> In the Victorian era, the home was the center of funeral rites. After death, two calls were made. One was to the doctor, who would come out to make sure the deceased was, indeed, deceased and not in a coma. The second was to the undertaker, who would come out to perform his services (embalming would have been done in the kitchen or a bedroom). . . . Superstition had its place, too, in the Victorian funeral process. A family would stop the clock at the exact time of death, then restart it after burial. And mirrors were covered with black material to keep the deceased's spirit from going into the mirror and remaining in the house."[6]

We shudder to think what it was like for both Lizzie and Emma to sleep upstairs with the bodies of their murdered parents below. While this was normal for the era, it must have been an eerie sight to see Andrew Borden laid out mere feet from where he had been obliterated with an axe.

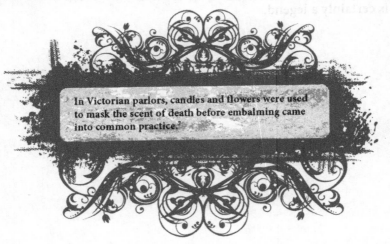

In Victorian parlors, candles and flowers were used to mask the scent of death before embalming came into common practice.[7]

What evidence was lost by allowing the body to remain at the Borden household? A modern detective would surely cringe at the thought of leaving the victims with the prime suspect so that she could hide or change evidence. Yet, like all her nineteenth century counterparts, Borden had no comprehension of DNA evidence. We may lament lost evidence, though the fact remains that Lizzie Borden was found not guilty by a jury of twelve men. She inherited her parent's money, finally allowing her a more financially comfortable living until her death in 1927.

An illustration of Lizzie Borden at trial.[8]

Borden died almost one hundred years ago but her legacy lives on in mythic proportions through the media adaptations of today. She is considered by most historians to have wielded the axe on that stifling, hot summer afternoon in a fit of homicidal rage that is still considered unique for the fairer sex. Lizzie Borden may not be a serial killer, but she is certainly a legend.

CHAPTER TWO
From Hell
(Jack the Ripper)

Despite the fact that we will never know his true name, Jack the Ripper is an icon that surpasses many modern-day serial killers. He is the archetypal boogie man; a shadow who lurks in the darkened alleys of our souls, glistening blade in hand. The creatures under our bed might not be real, but Jack the Ripper is. Having killed at least five women in the Whitechapel district of London in 1888, Jack the Ripper's violent and shocking treatment of his victims still resonates in popular culture today. At the time of his brutalities, the media found instant fascination with Jack, quickly turning him into a worldwide icon. He was considered the first murderer to be sensationalized in his time, though certainly not the last.

There were witnesses to the Ripper's crimes during his murder spree, but the police didn't commission an artist to sketch a likeness. Instead, they allowed a press artist to draw what they thought an evil murderer may look like![1]

One vital piece of the Jack the Ripper puzzle is understanding not only the era, but the community in which he thrived. Whitechapel was not the idyllic Victorian London we know from Charles Dickens's novels. It was where people, many immigrants, gathered to attempt a better life with no money, no prospects, and no notion of how truly frightening their neighborhood would become. About twenty years before Jack the Ripper made Whitechapel his hunting ground, author John Hollingshead wrote in his book *Ragged London* (1861) of his experience in the burg:

> Whitechapel may not be the worst of the many districts in this quarter, but it is undoubtedly bad enough. Taking the broad road from Aldgate Church to Old Whitechapel Church—a thoroughfare in some parts like the high street of an old-fashioned country town—you may pass on either side about twenty narrow avenues, leading to thousands of closely-packed nests, full to overflowing with dirt, misery, and rags.[2]

As the decades continued, things only got worse in Whitechapel. The Jewish and Irish immigrants were met with fierce discrimination, and the squalor of the area led to many women having no choice but to make money from sex work. Because of the dark, lonely streets and the desperation to survive, these women were highly vulnerable to Jack the Ripper.

This trend of serial killers choosing sex workers is an alarming one. Gary Ridgway, known as the Green River Killer, shared Jack the Ripper's victim profile one hundred years later; of his astounding forty-nine victims, the vast majority were sex workers and teen runaways. The Long Island Serial Killer, still unknown, preyed on women advertising sex on Craigslist. As we researched the sex work and murder connection, the statistics were staggering:

In a study conducted in Colorado Springs, USA, over four decades, researchers concluded that cis-gendered female sex workers (their sample was overwhelmingly made up of street-based sex workers), while they were actively working, were eighteen times more likely to be murdered than women of the same age and race from the general population. This estimate is based on just one geographical location so we cannot know if the same results would be found in other studies, although other estimates have tended to be higher. For example, it was estimated that female sex workers were sixty to one hundred and twenty times more likely to be murdered in Vancouver, Canada, than women from the general population. In a UK-based study of sex workers in London, cis-gendered female sex workers' mortality rate was recorded as twelve times higher than women from the general population and murder was identified as one of the leading causes of death.[3]

Why do many serial killers follow in Jack the Ripper's infamous footsteps? In his article for A&E, Adam Janos theorizes one reason is that the police "won't look as hard for a missing sex worker as they will for a more 'respectable' victim."[4] In an interview with Janos, social psychologist Eric Hickey concurs:

> Generally, police are not real fond of prostitutes, because there tends to be other kinds of crime going on when there's prostitution in the area. And so when someone goes missing, maybe one of their friends who is also a prostitute might go and report it . . . but usually, from the stories we hear, there are two or three prostitutes who disappear before they start to get a little nervous.[5]

Unfortunately, it seems that some of those working in law enforcement, as well as many community members, are often less concerned with the demise of sex workers than they are other members of our society. This concept of the "less dead" has been popularized recently. People are starting to notice that there was more public outcry over murderers who focused on college girls, like Ted Bundy, than the Gary Ridgways of the world.

In his article, "Serial Killers Prey on the Less Dead" for the *Seattle Post-Intelligencer*, reporter Mike Barber spoke to the parents of Tia Hicks. When Hicks went missing in 1990, the Seattle police purged Hicks's missing report from the system and did not investigate, supposedly because of her background as a possible sex worker, drug user, and runaway. Barber's article extrapolates on this trend:

> Criminologist Steven Egger calls the victims of serial killers "the less dead" because they are usually people who have been

marginalized—prostitutes, drug users, homosexuals, farm workers, hospital patients, and the elderly. "We don't spend a lot of time dealing with missing people who aren't particularly important; who don't have a lot of prestige," said Egger, a University of Houston-Clear Lake professor and former police officer. It's a public failing as well as a police failing, a common belief being that such people take big risks and get what they deserve."[6]

This unsettling look into our society is starkly depicted in one of the more popular films based on Jack the Ripper, *From Hell* (2001) starring Johnny Depp as Inspector Abberline and Heather Graham as sex worker Mary Kelly. While there is much speculation on the part of the filmmakers, and the writers who created the graphic novels that the film evolved from, both the Inspector and Kelly were based on real people. In fact, Abberline has been depicted numerous times in film, including by Michael Caine in *Jack the Ripper* (1988) and by Clive Russell in *Ripper Street* (2012). Frederick Abberline was, indeed, the Chief Inspector of the London Police at the time of Ripper's reign, although he was not clairvoyant

Aldgate Church in Whitechapel still stands today, although it has gone through several rehabilitations over the centuries. During one such renovation, the severed head of who is believed to be Henry Grey, First Duke of Suffolk, was found in the crypt. He had been executed in 1554 for treason.[7]

as depicted in *From Hell*. If he was, perhaps the culprit would've been found! Another fictional aspect of Abberline in the film is his addiction to opium. Abberline's drug use brings about visions of the murders, whether supernatural or hallucinatory. This led us to research what the reality of opium use was really like in Victorian London.

Virginia Berridge draws quite a vivid picture of the ubiquitous nature of opium in her article, "Victorian Opium Eating: Responses to Opium Use in Nineteenth Century England." "In the first half of the nineteenth century, opium preparations were freely on sale to anyone who wanted to buy them, in any sort of shop; they were carried about the countryside by hawkers, sold in grocers and general stores and on market stalls."[8] As the decades wore on, it became clear that opium use was becoming a health crisis in England. Berridge describes the fallout:

A growing official uneasiness about opiate use did develop, however, and eventually found expression in the restrictions of the 1868 Pharmacy Act and in changed attitudes toward the drug. Statistics on opiate deaths caused concern too—the publication of coroner's returns of deaths by poisoning in England and Wales in 1839 revealed that one hundred and eighty-six out of a total of five hundred and forty-three such deaths were the result of opium poisoning. The Registrar General's Office collected scattered figures for opiate deaths which were published in the late thirties: in the two years 1838 and 1839, there were twenty-seven opiate deaths in London out of a total of one hundred and twenty-five poisonings. The 1840 report revealed that five deaths per million living were the result of opium poisoning. The first series of figures on opium beginning in the early 1860s showed the full extent of the situation: one hundred and twenty-six deaths from opiates in 1863, for instance, out of a total of four hundred and three poisoning fatalities, with eighty deaths in that year and ninety-five in 1864 from laudanum and syrup of poppies alone. Around a third of all poisoning deaths in the decade were the result of the administration of opiates, and the relatively high accidental, rather than suicidal, death rate from opiates bore witness to the drug's easy availability.[9]

All of this occurred decades before Jack the Ripper and Inspector Abberline played cat and mouse in the 1880s. As depicted in the film, "riding the dragon," or getting high on opium, was not considered as socially acceptable as it once was. Despite this, opium dens prevailed, especially in places as corrupt and dangerous as Whitechapel. Reporting on an opium den in Whitechapel in 1868, the French journal *Figaro* describes, "It is a wretched hole . . . so low that we are unable to stand upright. Lying pell-mell on a mattress placed on the ground are Chinamen, Lascars, and a few English blackguards who have imbibed a taste for opium.[10]

So, what, precisely, does opium do to a user's body? In *From Hell*, Johnny Depp's Abberline lies almost coma like, reveling in the relaxing properties of the drug derived from a poppy flower. Because about 15 percent of the alkaloids, or organic compounds, in opium are morphine, it causes both a "high" and dangerous side effects including cardiac arrest, lung edema, and respiratory collapse. Although the users in Victorian England didn't understand its full potency, small amounts of opium can be useful for chronic health conditions if used under the direction of a physician. Thankfully, unlike the fictional Abberline in *From Hell* who died at a young age because of his opium addiction, the real Inspector refrained from such recreational drug use and lived to the ripe old age of eighty-nine.

Mary Kelly is considered the last known victim of Jack the Ripper. She had been married at age sixteen, losing her husband in a mining accident a few years later. Destitute, she became a sex worker with no other prospects on her horizon. Unlike the other victims, Kelly was killed indoors, and was mutilated the most profoundly after her death. This, of course, is quite dramatic in the film, as we, the viewers, have been following her journey to uncover the Ripper.

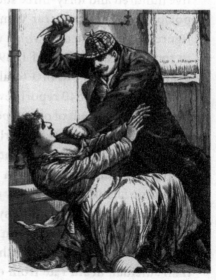

Jack the Ripper was never identified.

One important achievement of *From Hell* is the humanization of the Ripper's victims. Most serial killer movies based on reality tend to focus almost exclusively on the killers themselves. Because Jack the Ripper is never identified, and shown only in shadow and suggestion, this leaves room to empathize with not only Abberline, but most vitally, the women whose lives were taken.

The beginning sequence of *From Hell* begins with a look into the cramped, putrid life of those living in Whitechapel. Sex and drunkenness are on full display below the dimly lit businesses, an ideal spot for a man to go undetected. Jack the Ripper is not the only danger to his future victims; Mary Ann Nichols, Annie Chapman, Elizabeth Stride, Catherine Eddowes, and Mary Jane Kelly. Because these women were not bred in high society, they struggle to make enough money to even eat, much less thrive. The movie depicts the reality of gangs, misogyny, xenophobia, rape, and more. While there is speculation about these women's true personalities and dreams, by watching *From Hell*, we can begin to understand the suffocating world they lived in.

Jack the Ripper's shadowy presence still haunts the streets of Whitechapel. For twelve British pounds, you can join in on the Jack the Ripper Tour, a narrated walking group that meets at night. Highlights

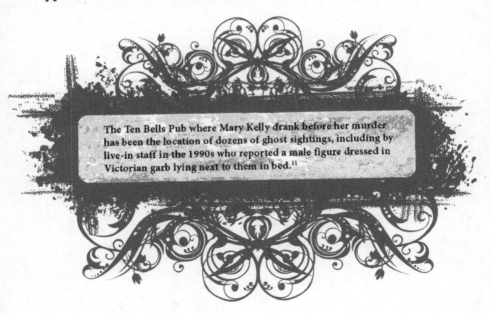

The Ten Bells Pub where Mary Kelly drank before her murder has been the location of dozens of ghost sightings, including by live-in staff in the 1990s who reported a male figure dressed in Victorian garb lying next to them in bed.[11]

include "A warren of atmospheric old streets that have hardly changed since they formed the backcloth against which the Jack the Ripper saga was played out."[12] The prime suspects are also discussed. Was it John Pizer, a bootmaker known for assaulting sex workers? Could it be Dr. Thomas Neill Cream, a physician who would have certainly known how to dissect the bodies with such precision? Whoever it may have been, *From Hell* reminds us that Jack the Ripper's story is more complex than just one monster stalking the streets. It is a story of the haves and have-nots, of women who were vulnerable victims simply because of their place in society.

CHAPTER THREE

The Devil in the White City (H. H. Holmes)

"America's first serial killer" is quite a foreboding moniker. H. H. Holmes wears that hat well. Unlike the slathering, depraved boogeymen who women and children feared would follow them down a dark alley, Dr. Holmes was the sort of gentleman mothers hoped their daughters would marry. Of course, his edifice of charm and ambition was exactly that: a weak mask that eventually slipped from the doctor's face.

While many serial killers have been given the cinematic treat-

Serial killer H. H. Holmes.

ment, H. H. Holmes has found a resurgence of interest because of Erik Larson's bestselling book *The Devil in the White City: Murder, Magic, and Madness at the Fair that Changed America* (2002). Nearly twenty years after its publication, the novel-style nonfiction account of H. H. Holmes's depravity amid the Chicago World Fair is still a popular read. Currently, director Martin Scorsese and producer Leonardo DiCaprio are working to bring a television adaptation of Larson's book to Hulu. Until then, H. H. Holmes will have to wait for his Hollywood treatment.

Because it has been well over a century since his murderous reign, the facts are fuzzy. This is a particular case where folklore and history blend. What we do know is that after receiving his medical degree from

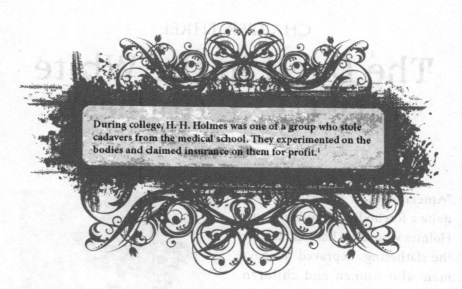

During college, H. H. Holmes was one of a group who stole cadavers from the medical school. They experimented on the bodies and claimed insurance on them for profit.[1]

Michigan State University, Herman Mudgett thought it best to change his name. He was a young doctor, headed to the prosperous city of Chicago, and thus needed a bit of an ego boost. The name H. H. Holmes had the ring he longed for, a name that he felt fit his grand, new life. It was also a name that would become notorious across the United States, one synonymous with the likes of Jack the Ripper.

Legend has it, though there is no proof, Holmes killed a pharmacist in order to obtain their drug shop, which was the catalyst for what would become Holmes's "murder castle." Just a few miles from the center of the Chicago World Fair, Holmes hired workers to expand on the pharmacy, developing it first into a sort of strip mall, and then into a structure designed to secretly torture and kill its inhabitants. In her article, "American Gothic: The Strange Life of H. H. Holmes," Debra Pawlak describes the labyrinthine hotel:

Mazes of mystery entwined the second and third floors of Holmes's castle. There were secret hallways and closets connecting the seventy-one bedrooms. Soundproof, and with doors that could only be locked from the outside, these "guest quarters" were fitted with gas pipes attached to a control panel in Holmes's bedroom. He turned them on and off at will. Holmes's office, complete with an oversized stove, was also on the third floor adjacent to his walk-in vault. There were trap doors, sliding panels, stairs that led nowhere, and doors that opened to nothing but solid brick walls.[2]

Unbelievably, like the fictional "Demon Barber of Fleet Street," Sweeney Todd, there was purposeful rigging to move the deceased. "Large, greased chutes led straight to the basement where Holmes kept an acid tank, a dissecting table . . . and a crematorium."[3]

While many serial killers are defined by their specific aesthetic in chosen victim (think Ted Bundy's penchant for college co-eds with long, brown hair), Holmes seemed to murder anyone who stood in the way of his burgeoning ego. This included his mistress Julia Smythe and her daughter Pearl, who both vanished under Holmes's care after gossip that Julia had become pregnant with his child. Most disturbingly are his murders of Benjamin Pietzel and three of his children. Holmes convinced Pietzel that they could scam an insurance company by faking Pietzel's death with a cadaver, but Holmes ultimately burned him to death, then murdered his kids Alice, Nellie, and Howard to obscure his misdeeds. His annihilation of this family ultimately led to Holmes's capture and hanging in 1896.

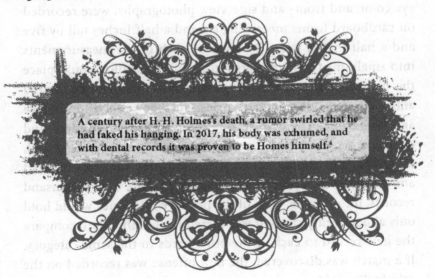

A century after H. H. Holmes's death, a rumor swirled that he had faked his hanging. In 2017, his body was exhumed, and with dental records it was proven to be Homes himself.[4]

Unlike many of Holmes's supposed crimes, the slayings of the children were proven with irrefutable evidence. "Alice and Nellie's bodies were found in a Toronto cellar. Later, authorities found teeth and pieces of bone among charred ruins that belonged to Howard in an Indianapolis cottage that Holmes had rented."[5] While I was watching a documentary on Holmes, *H. H. Holmes: America's First Serial Killer* (2008), the term "Bertillon method" was used to describe how the investigators identified the bodies of Alice, Nellie, and their burned father, Benjamin in the late nineteenth century.

Developed by French police officer Alphonse Bertillon, this method was based in anthropometry, essentially the measurement of the human form. Before Bertillon, the only way for law enforcement to identify remains was through photographs which were rare and not particularly high quality. Raised in a family of statisticians, Bertillon believed that there had to be a better, more scientific way to discover a found body's identity.

Bertillon took measurements of certain bony portions of the body, among them the skull width, foot length, cubit, trunk, and left middle finger. These measurements, along with hair color, eye color, and front- and side-view photographs, were recorded on cardboard forms measuring six and a half inches tall by five and a half inches wide. By dividing each of the measurements into small, medium, and large groupings, Bertillon could place the dimensions of any single person into one of two hundred and forty-three distinct categories. Further subdivision by eye and hair color provided for one thousand seven hundred and one separate groupings. Upon arrest, a criminal was measured, described, and photographed. The completed card was indexed and placed in the appropriate category. In a file of five thousand records, for example, each of the primary categories would hold only about twenty cards. It was therefore not difficult to compare the new record to each of the other cards in the same category. If a match was discovered, the new offense was recorded on the criminal's card.[6]

Even Sherlock Holmes was a fan of this newfangled mathematical criminology, as referenced in Sir Arthur Conan Doyle's *The Hound of the Baskervilles* (1901), as Detective Holmes is considered the "second highest expert in Europe."[7]

The Bertillon Method came to the US in 1887, first used in the Illinois State Penitentiary.[8]

While fiction authors like Sir Arthur Conan Doyle have long been influenced by true crime plaguing the news, it is no surprise that the wildly violent H. H. Holmes evokes the same creative interest. We had the great fortune to talk to horror writer Sara Tantlinger. Her book of poetry, *The Devil's Dreamland: Poetry Inspired by H. H. Holmes*, won the 2018 Bram Stoker Award for Best Poetry Collection.

Meg: "Could you tell us what drew you to H. H. Holmes as a creative?"

Sara Tantlinger: "I watched a documentary on Holmes a few years ago and was just fascinated. He's surrounded by so many theories. Obviously, forensic science in the 1800s wasn't quite capable of doing everything it can currently, so while we may speculate and have some evidence here and there about what Holmes did, how many victims he had, and what his motives were, the truth is, we will never know for sure. As a writer, that stark gray area invited me in to explore the twisted possibilities. I could take fact and imagination and create something entirely my own."

Kelly: "In the author's note of *The Devil's Dreamland*, you point out that 'this is not a history lesson' and 'both fact and speculation intertwine.' Why was it important for you to say this to your readers?"

Sara Tantlinger: "Great question! I wanted my readers to know that, as a work of historical horror, not everything I wrote about was derived from what we believe may be 'facts' surrounding Holmes. But, at the same time, I completed such an intense amount of research. It was important for me to let readers know how dedicated I was to understanding anything I could about a man who was, frankly, impossible to understand or know."

Meg: "Your point of view alters from victims and worried family members to wives of Dr. Holmes and, of course, the killer himself. Which was the most emotional for you to inhabit? What was it like getting inside such a sadistic man's head?"

Sara Tantlinger: "The most emotional for me was thinking (and reading) about the many women whose entire lives he ruined. While he did not kill any of his three wives, he did murder mistresses and other women whom he charmed into giving him money, usually before killing them. These women had families who were searching desperately for them, hoping for a safe return, and simply not knowing why their loved ones had disappeared. I used great care when writing from their points of view and tried to give them voices and agency throughout their poems. Getting inside Holmes's head was an interesting process. His prison diaries and memoirs are available through the Library of Congress, so I read through those a few times, which helped me see how articulate Holmes was. He wrote lies in such an idyllic way . . . and focusing on that snakelike charm helped me see how he convinced his victims to do what they did. I had to focus on it heavily and didn't really go out or communicate a lot with my friends for a few months because the research coupled with my dark imagination running wild created an atmosphere I knew I needed to stay entrenched in to complete the work."

Kelly: "In your poem, 'Holmes vs. The Ripper, Part II' you imagine what H. H. Holmes would make of the news stories of Jack the Ripper from across the pond. Do you agree with Holmes's assessment, that he had a different motive and method? Or do you find them quite similar?"

Sara Tantlinger: "While I think Holmes and the Ripper might have found a great deal to talk about had they ever met, I wanted to do the 'Holmes vs. The Ripper' poems in particular because of how much I disagree with the theory that they could have been the same person. Even if it was logistically possible in regard to time periods and locations, the Ripper was grossly intimate with his victims and very specific with his targets. Holmes was more of a coward until the person became only a body, and then he seemed more able to strip the skeleton of its flesh. His motives seemed to be entirely for monetary gain or to get rid of an inconvenience. I haven't studied the Ripper enough to know everything about him, but it seems he received more . . . enjoyment from his actions compared to Holmes."

Meg: "Anything else you'd like to add about your journey writing this book?"
Sara Tantlinger: "One more thing that is personally important for me to add is that with Holmes, since everything happened so long ago and no one is currently suffering from his atrocious actions, it put a kind of distance between myself and the writing that allowed me to feel okay with authoring what I did. I do get concerned when current serial killers are sensationalized (for instance, the many Ted Bundy movies that are out) because there are victims and families still suffering from those actions. So, I would ask anyone thinking of writing about historical horror or putting real-life serial killers into their fiction to . . . make sure no one is profiting off of anyone's current pain, and to be thoughtful overall with the writing."

Kelly: "Tell our readers about current and upcoming projects of yours!"
Sara Tantlinger: "Thank you so much! In 2020, I released my third poetry collection, *Cradleland of Parasites,* which draws inspiration from the Black Death, and an anthology I edited titled *Not All Monsters,* stories by women in horror. Both books are published by Strangehouse Books and are available on Amazon!"

Thank you to Sara Tantlinger for the incredible insight into H. H. Holmes. We particularly liked Sara's poignant thoughts on the importance of keeping the victims in mind, whether focusing on the fictional films and books, or the true, grisly facts.

SECTION TWO
FOREIGN MURDER

CHAPTER FOUR
Lake Bodom (Unsolved)

I (Kelly) spent eleven summers of my childhood attending a foreign language camp through Concordia Language Villages. Salolampi Finnish Language Village in Minnesota was a magical place where I connected with fellow kids of Finnish descent and learned to speak the language of my ancestors. We learned folk dancing, ate authentic Finnish food, and sang songs that I can remember to this day. When I traveled to Finland in high school to meet relatives, I was impressed with the culture, environmentalism, and overall kindness of the country, and was shocked to learn that *Lake Bodom* (2016) was based on a horrific true crime that took place there in 1960. The film offers an updated take on what may have happened that night.

In June 1960, four teens went camping by Lake Bodom near the city of Espoo, Finland. Sometime between the hours of 4:00 and 6:00 a.m., a killer attacked them from outside of their tent. Three of them were murdered by stab wounds and blunt force trauma while the fourth teen was found outside the tent with stab wounds and

The 1960 crime scene near Lake Bodom in Finland.

other injuries. He survived and was charged for the crime forty-four years later but was never convicted. The murder remains a mystery still and continues to be the most famous unsolved murder in Finland's history.

In the movie, a group of four friends drive out to Lake Bodom to reconstruct the crime area in order to test a theory. Is this method used in real investigations? Absolutely. Criminal investigations use this method to test out theories and better understand a crime scene. There are certain steps to follow to get the best results. First is the recognition of evidence. Second is the documentation, collection, and evaluation of evidence. This is followed by forming a hypothesis, testing it, then reconstruction. The final step is reporting the findings and determining if the crime likely occurred in any given manner.[1] This information can be used to profile the criminal, weapon, or the method of entry along with other aspects of the crime.

Nearly one million crimes or violations (minor offenses) are reported to the police annually in Finland. About half of them are traffic violations.[2] In the United States, property crimes outnumbered violent crimes in 2018, numbering 7.2 million and 1.2 million respectively. Larceny was the most common property crime with 5.2 million incidents, while aggravated assaults accounted for around two thirds of violent crimes.[3]

Two of the film's characters go swimming in the lake at night. I took many late-night swims in my childhood at our cabin. There was one lake, though, that I never went in at night. It was called Spirit Lake, better known as Dead Man's Lake. My parents told the story of a man who drowned there decades earlier . . . and his body was never found. The lake was named after this incident and locals would call it "Deady's." The truth is, there are numerous examples of bodies never being recovered from lakes. According to Dr. Anton Sohn, chairman of the pathology department at the University of Nevada in Reno, many factors contribute to drowning victims never being found:

> When people drown . . . their lungs fill with water, dropping them into the depths of the lake. Death brings decomposition, where bacteria consumes bodily flesh at some pace. During that process, gases such as methane, nitrogen, and oxygen are produced but the type of gases formed depend on the type of bacteria in the gastrointestinal tract . . . the gases would allow a body to rise. Since the lake has frigid temperatures, bodies don't decompose, thus gases don't form, prompting them to stay submerged.[4]

The lake in the film plays a critical role and adds a chilling plot twist to a seemingly predictable story.

In the Lake Bodom murder investigation, police did not cordon off the site nor record the details of the scene (later seen as a major error) and almost immediately allowed a crowd of police officers and other people to trample around and disturb the evidence. The mistake was further exacerbated by calling in soldiers to assist with the search around the lake for the missing items, several of which were never found.[5]

We had the opportunity to interview Taneli Mustonen, the writer and director of *Lake Bodom*, to get an insight into the filmmaking process.

Kelly: "What drew you to this story and what made you decide to write it in this fictionalized, updated way?"

Taneli Mustonen: "In short? Everything. Literally, this case has all the components of a great murder mystery or awesome horror film. We must remember that Finland at the time (1960) was a very different country—economy was booming after World War II, uniting people that now could plan and build better lives for themselves and their families. Finland was also very safe, it wasn't unusual that parents could send their kids out alone in bigger cities or even across the country to meet relatives during summer holidays. Now, all that quickly ended when Bodom happened. It was and still is to this day such a heinous act of nameless, senseless violence against innocent teen campers that the waves of this shock can still be felt whenever Bodom is mentioned."

Kelly: "I was surprised to hear about the murder because, in my mind, Finland has always been so safe and crime free!"

Taneli Mustonen: "For me, like for so many, the case of Bodom is something you learned the first time you asked permission from your parents to go out camping with your friends. 'You can . . . but remember what happened to those kids at Lake Bodom.' Maybe you went camping anyway, like me, and that story became the *thing* you scared your friends with, adding bits and pieces here and there. So, the Bodom case has grown with every generation, news story, and investigation. The place is still there, the murder case still unsolved, but there's also so much made up on top of it. And that is also part of the fascination; every generation rewrites the history of Bodom."

Meg: "I remember as a child my parents making it seem like there was a serial killer around every corner!"

Taneli Mustonen: "Naturally, I wanted to make this a horror film. But even as a kid in my film school days I knew that *everyone* who shared my passion for horror had the same idea. Bodom just was this perfect slasher template. So, I got frustrated and one time rode to the lake. And there I saw these kids, wandering on the rocks and making notes. At first, I thought they were there just to goof around, but then realized that they were making some kind of reconstruction of the actual happenings. I went back to film school and told this to my friend (now the cowriter and producer of *Lake Bodom)* Aleksi Hyvärinen. He got so excited and said that this could give this story something new and fresh. And from there (almost exactly ten years ago) we had the film in our hands.

Kelly: "That's incredible! It always surprises me how long the filmmaking process can take. How did the process continue from there?"

Taneli Mustonen: "The reconstruction of the murder case wasn't the end but the mere beginning. We also did our research and met this former homicide investigator who told us something that really resonated with what we wanted to say. He said that most of the murders he ever investigated could be traced back to love. Yes, *love.* He said that nothing is more evil than love that is denied or unanswered in return. This literally blew us away. And with that lightning bolt of an idea, we ran back to write the whole story once more. And oddly enough, with this we ended up telling something real about the original murder case. Most of the investigations point in this direction that maybe the murderer is in fact one of the victims. Maybe he too was denied love in return? And so, the story that we took from past to present with these new kids also became a way for us to say something we believed about this case."

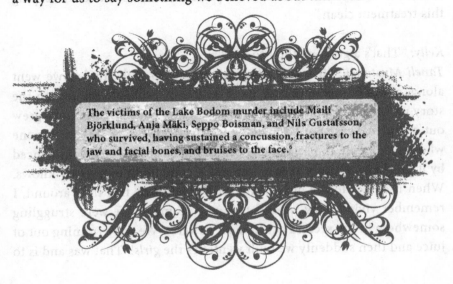

The victims of the Lake Bodom murder include Maili Björklund, Anja Mäki, Seppo Boisman, and Nils Gustafsson, who survived, having sustained a concussion, fractures to the jaw and facial bones, and bruises to the face.[6]

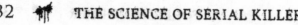

Meg: "We appreciate how you subverted the expectations of the audience by having the two seemingly innocent girls actually be the initial perpetrators of this crime. Do you have any speculations about what actually happened that night in 1960? How deep into the research did you get?"

Taneli Mustonen: "The writing process of *Lake Bodom* was very fluid. It consisted of us first really just coming up with a story that we felt would inspire audiences—even without any connection to the actual case of Bodom. It may sound a bit strange, but we felt it was necessary. We needed to find *our* story, what would inspire us. So, like always for us, we took a day to ourselves, with a goal of getting something concrete on paper. No calls, interruptions, or pressure—just chatting—and dreaming, really. That day turned into two really fruitful ones. *Lake Bodom* actually showed us that creative limitations are very important for new writers like us at that time. We decided that we would only have four teenagers, two girls and two boys, camping for one night in a remote forest (this later changed but that was the initial plan). What could we come up with? What are the dynamics, relations? And that spurred quite quickly into what became *Lake Bodom*. Another thing was that we wanted to have a short treatment, no more than one page of synopsis, but something that would have the whole story. And to avoid us cheating ourselves with fancy language or anything like that, we used this rule that we only use two lines for any scene. I know it may sound silly, but that forced us to really concentrate on a) what takes place in the scene, and b) keeping this treatment clean."

Kelly: "That's a great method to use!"

Taneli Mustonen: "It was really just coming up with stuff as we went along, a totally new thing for us. And that is probably the reason the story unfolds like it does. It is hard to tell how much of our story grew out of the real case but I'd imagine it played a big part, like for anyone who researches it. In Finland it is almost impossible not to be influenced by it. Like I said, the real case was and is something everybody knows. When it comes to the girls being the perpetrators this time around, I remember vividly how we had this *omg* moment. We were struggling somewhere in the second act (like all writers). We were running out of juice and then suddenly we both went: 'It's the *girls*!' That was and is to this day one of the most exciting moments that has happened to us. And with that we could link the story to the original case in a way that doesn't point any fingers of blame or accuse someone without any evidence and also make our story stand all on its own. Because it was a real worry that we had. The last thing we wanted to do was hurt any relatives or families of the victims. In fact, we were shooting this other series at the time, and when we mentioned that we were working on this film about Bodom, one of the actors said that he was a family member of one of the girls who died that horrible night. So, that was a wakeup call for us as well. It is another thing to make a film and then carry that weight of hurting someone with such a personal loss."

Meg: "That's so important and it's not always present in movies about true crime."

Taneli Mustonen: "We really wanted to avoid becoming blamers or accusers. We had and still have no idea who was behind these grim murders. Naturally, we did our research, even to the point of finding these crazy theories (thanks to the internet) that are just laughable, really. But to make that shift from filmmakers to jury—that was something we wanted to avoid at all costs. Of course, we have our own idea of what really happened. We felt and still do, that there is grounds for a case against one of the boys, the one that survived the attack. He has now passed on, and the latest trials in the 2000s were mainly about getting him to confess, but as it didn't happen and there just was not enough actual hard evidence against him, he was set free. Bodom is still very much an open case, open wound really, but how we see our film is that in it we do give this hint of what could have happened on that tragic night. That's plenty enough for us as filmmakers. And we can have our minds at peace."

Kelly: "Absolutely!"

It was fascinating to learn more from Taneli Mustonen about the creative process of making a film and it opened our eyes to the complex layers of the art. Check out these other Finnish horror movies: *The White Reindeer* (1952), *Sauna* (2008), and *The Moonlight Sonata* (1988).

CHAPTER FIVE

Rillington Place
(John Christie)

John Christie murdered at least eight people, including his wife Ethel, during the late 1940s and early 1950s in London. Most of the victims' bodies were discovered behind the walls and underneath the floorboards of his home located at 10 Rillington Place. Based on this true story, the 2016 miniseries *Rillington Place* aired on the BBC and stars Tim Roth as Christie.

The miniseries begins with Timothy Evans (Nico Mirallegro) being convicted and hanged for the murder of his wife and infant daughter. As he is led to his death, he pleads that it's Christie who is the guilty party. The sad story behind this moment is that it's absolutely true. Evans, his wife Beryl, and their daughter Geraldine lived upstairs from the Christies. Little did they know that they would all become victims of this brutal murderer. Christie's first two victims, twenty-one-year-old Ruth Fuerst and thirty-two-year-old Muriel Eady, were buried in the garden. Next, he disposed of twenty-year-old Beryl Evans in a sewer drain. Twenty-five-year-old Rita Nelson, twenty-six-year-old Kathleen Maloney, and twenty-six-year-old Hectorina McLennan were later discovered in the walls of the kitchen. The stench of the rotting bodies hidden within the walls eventually became so overpowering that he moved from the house. When the new residents renovated the kitchen, they discovered the source of the gruesome smell and also recovered the body of Ethel Christie from under the floorboards in the parlor.[1]

The doctor that Ethel talks to about her husband says that middle-aged men need extra care and attention. Instead of listening to her concerns, he encourages her to be better for her husband. Clearly, her husband is depressed and has other issues, but the doctor blames Ethel for his melancholy. In a 2016 interview, Samantha Morton was asked about

"The police made several mistakes in their handling of the case, especially in overlooking the remains of Christie's previous murder victims in the garden at Rillington Place; one femur was later found propping up a fence."[2]

how her character, Ethel, is viewed as a symbol of humans' capacity to love or at least accept monsters:

> I find it absolutely fascinating, the psychological aspect of love, how far love will go and what women in abusive relationships will often do for their partner, like protect them in court. You hear a lot of horrific cases of child abuse where the woman will support the father to the detriment or even the death of their child. I cannot get my head around that, but then I've never been in that position. So, nothing on a personal level, but it really developed my acting chops. It was a challenge to play someone so submissive.[3]

Historically, women have often been dismissed by doctors. In a report by *Today*, physicians who are fighting to change the system said that "it's not malice but a pervasive, implicit sex and gender bias in medicine that's leading female patients to be misdiagnosed, neglected, dismissed as complainers, accused of being overanxious, mislabeled as depressed, or told their symptoms are all in their heads. . . ."[4]

Christie offers to help Beryl with an abortion in the miniseries. What are the statistics about abortions? Between 2015 and 2019, on average,

73.3 million induced (safe and unsafe) abortions occurred worldwide each year.[5] Of those, 45 percent were considered unsafe. Women may choose these unsafe options because they don't have access to legal abortion, have an untrained provider, or may have an abortion with unsanitary tools or surroundings. "Some sixty-eight thousand women die of unsafe abortion annually, making it one of the leading causes of maternal mortality (13 percent). Of the women who survive unsafe abortion, 5 million will suffer long-term health complications."[6] In the miniseries, Beryl is dead when her husband comes home. Christie claimed she was septic and there was no way to save her but it's later revealed that he strangled her.

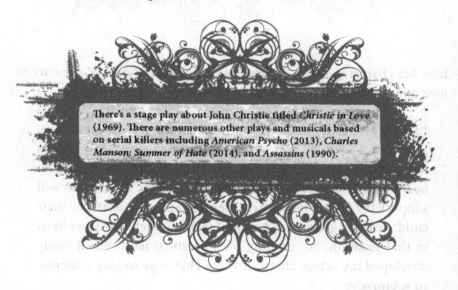

There's a stage play about John Christie titled *Christie in Love* (1969). There are numerous other plays and musicals based on serial killers including *American Psycho* (2013), *Charles Manson: Summer of Hate* (2014), and *Assassins* (1990).

Christie was considered to be a necrophile. Necrophilia is a pathological fascination with dead bodies, which often takes the form of a desire to engage with them in sexual activities. Incidents have taken place throughout history and a disturbingly large number of serial killers take part in these activities.

Christie used strangulation to kill his victims. Strangulation is "squeezing of the neck with enough force to block the flow of blood to the brain and/or the flow of air to the lungs. The loss of blood flow deprives the brain cells of vital oxygen. Even short periods of time without oxygen can cause damage to the brain."[7] This loss of oxygen to the brain may cause confusion, memory problems, loss of consciousness, or even death. There are other possible causes of oxygen deprivation to the brain including stroke, carbon monoxide poisoning, and brain injury.

John Christie's trial lasted only four days, and the jury returned a verdict of guilty after deliberating for only an hour and twenty minutes. Christie was sentenced to death and hanged just over two weeks later at Pentonville Prison in London on July 15, 1953.[8]

The bodies hidden under the floorboards and in the walls of Christie's home were covered in lye to hasten decomposition. Does this actually work? Criminals have been known to use sodium hydroxide or potassium hydroxide, strong bases commonly known as lye. "Heated to three hundred degrees, a lye solution can turn a body into tan

Human bodies naturally decompose beginning twenty-four to seventy-two hours after death and will begin to liquify after a month.[9]

liquid with the consistency of mineral oil in just three hours."[10] Bones of the victims were found and identified. What is the science of identifying bones of humans versus animals? It all begins with the skull. Human skulls have a larger cranial vault to hold our brains and are oriented on a vertical axis while non-human animal crania are oriented on a horizontal axis.[11] Differences in limbs, and how we use them, are also easily distinguishable for experts.

Some of the victims in the Christie case were identified by their hair. This process involves testing hair samples "to determine the color, shape and chemical composition of the hair, and often the race of the source individual. The presence of toxins, dyes, and hair treatments are noted. If the hair still has a follicle (root) attached, DNA testing may be used to identify an individual."[12] Christie collected the hair of his victims, but not all of the hair was identified, so it's been speculated that he was responsible for more murders than he was convicted of. Professor Keith Simpson, one of the pathologists involved in the forensic examination of Christie's victims, speculates:

> It seems odd that Christie should have said hair came from the bodies in the alcove if in fact it had come from those now reduced to skeletons; not very likely that in his last four murders the only trophy he took was from the one woman with whom he did not have peri-mortal sexual intercourse; and even more odd that one of his trophies had definitely not come from any of the unfortunate women known to have been involved.[13]

Christie claims in the miniseries that he was trained to be a doctor. Because of this claim, Beryl Evans trusted him with her abortion, which led to her death. Several of his other victims were given vapors to "cure" their bronchitis. Although he wasn't really trying to cure anyone, how has science changed regarding the treatment of bronchitis? Bronchitis has been a recognized medical condition for centuries and was named in 1808. Treatments of the past included garlic, pepper, cinnamon, and turpentine. Some later therapies emphasized coffee, ipecac, and potassium nitrate. Today, bronchitis is most often treated with rest, fluids, and non-steroidal anti-inflammatory drugs. Some people may find relief with inhaled corticosteroids to reduce airway inflammation.

Rillington Place was renamed in May of 1954 to Runton Close to get away from the association with the Christie murders. The place was eventually demolished in 1970 but the memories of the atrocities committed there will forever be remembered.

CHAPTER SIX
Wolf Creek (Ivan Milat)

"The following is based on actual events. Thirty thousand people are reported missing in Australia every year. Ninety percent are found within a month. Some are never seen again." So begins the movie *Wolf Creek*, released in 2005. Based on the Australia backpacker murders that took place between 1989 and 1993, *Wolf Creek* follows a group of friends out on an adventure that ends in tragedy.

Up until the mid-nineties, crime was considered rare in Australia. "Violent crime in Australia is relatively low, with fewer than one hundred reported cases of armed robbery, murder, or sexual assault per one hundred thousand persons nationally."[1] The three friends in the film depart on a road trip across Australia's outback. While researching road tripping in the outback, it was fascinating to learn some tips from experienced travelers. First, the roads are bumpy and you're able to "do laundry" while driving if you have a container with a lid filled with water and detergent. When you reach your destination, you can rinse and hang your clothes to dry. Second, it's important to be aware that cell phones will not work in the outback. A GPS or standard map is required. Third, many of the roads are unpaved, making them rough, so it's important to be aware of the stress that will be put upon your vehicle. Finally, it's important to fill up on fuel and other essentials whenever the possibility is presented. There are stretches in the outback when fuel and food are not available so having them on hand is important.[2]

The Wolf Creek meteorite site plays a role in the film as a tourist destination. This is where the killer disarms something on each victim's vehicle so that he can be the "savior" that comes in to save the day. Instead, he tows the unknowing victims to his abandoned outpost to torture and murder them. The meteorite site is described by Ben (Nathan Phillips) in the film as the equivalent of two hundred nuclear bombs going off at the same time. What kind of impact do meteorites have? It's interesting to note that "impacts of large meteorites have never been observed by humans. Much of our knowledge about what happens next must come from scaled experiments. As the solid object plows into the Earth, it will compress the rocks to form a depression and cause a jet of fragmented rock and dust to be

There are only one hundred and twenty-eight confirmed crater sites on Earth.[3]

expelled into the atmosphere."[4] Experts postulate that numerous events could be caused by meteorites including earthquakes, massive dust in the atmosphere like a nuclear winter, wildfires, tsunamis, and a change in the atmosphere. Only a few small meteorites have been witnessed throughout history, and while their impacts have been minimal, it is suggested that a large enough meteorite could cause mass extinction.

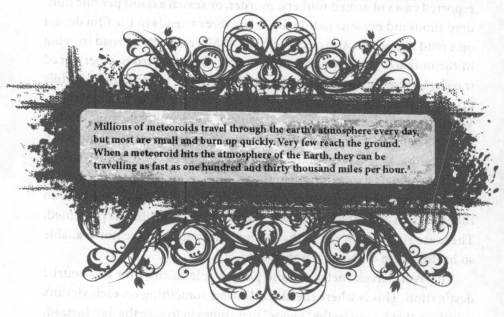

Millions of meteoroids travel through the earth's atmosphere every day, but most are small and burn up quickly. Very few reach the ground. When a meteoroid hits the atmosphere of the Earth, they can be travelling as fast as one hundred and thirty thousand miles per hour.[5]

Australia's outback is described as having towns that are covered in dust and abandoned. Is this true? There are indeed numerous ghost towns throughout Australia. Some of them have been preserved while others have been left to the elements. Many are mining towns while others are former penal colonies. The group in the film can easily see the stars while sleeping under them in the outback. Is this possible everywhere? A little over one hundred years ago, you could step outside and see the stars anywhere in the world. Due to light pollution, an excessive amount of artificial light, it's nearly impossible to see the stars from most highly populated areas today. The best places to view the stars include remote islands, national parks, and deserts.

Hooks and chains are hung in a makeshift shop in the movie akin to *The Texas Chainsaw Massacre* (1974). Mick (John Jarratt) mocks Kristy (Kestie Morrassi) with a gun before firing and missing her, purposely. What are the laws regarding guns in Australia? Following several high-profile killing sprees, the federal government in Australia coordinated more restrictive firearms legislation within all state governments. "In two nationwide, federally funded gun buybacks, plus large-scale voluntary surrenders, and state gun amnesties . . . Australia collected and destroyed more than a million firearms, perhaps one-third of the national stock." People can own guns but must demonstrate a genuine reason, not including self-defense, in order to get a license.

Dental records were used for identification in some of the real-life murders committed by Ivan Milat. What is the science behind this process? Tooth enamel is harder than any other substance in the body and will remain long after other parts of the body have decayed. "To identify a person from his or her teeth, a forensic dentist must have a dental record or records from the deceased person's dentist. In the case of an incident involving multiple deaths, forensic dentists receive a list of possible individuals and compare available records with the teeth and find a match."[6] This science can be used for other reasons too. DNA can be retrieved by extracting the pulp from the center of the tooth. Pulp can last for hundreds of years and has offered clues to scientists investigating historic events including the bubonic plague and other blood-borne diseases.

Ivan Milat stabbed his victims multiple times, often causing injuries that resulted in paralysis. What is the science behind paralysis? BrainAndSpinalCord.org explains:

When a victim is stabbed in the area of the spinal cord, the spinal cord can be severed, sheared, torn, or otherwise damaged. This will result in a loss of function below the point of injury. The extent of the injury, as well as where the injury occurred, will determine whether the injury is complete or incomplete. Complete injuries result in total loss of function, while incomplete injuries result in some degree of function loss.[7]

The gruesome injuries in Milat's cases also had another thing in common. "The victims were all young, they were all found covered in branches and leaf litter, and each had been 'attacked savagely, with a great deal more force than was necessary to cause death, and apparently for some form of psychological gratification.'"[8]

Link analysis technology was used to help catch the killer by tracking things like vehicle records and gym memberships. What is link analysis technology and how is it used? "Link analysis helps an investigator visualize complex 'links' within an investigation that may be hard to communicate with others or hard to discern. It's helpful in quickly visualizing outliers for a target-centric investigation and it enables you to put information on a graph. . . ."[9] In the Milat case, tracking potential victims' whereabouts narrowed the list of suspects from two hundred and thirty down to thirty-two.

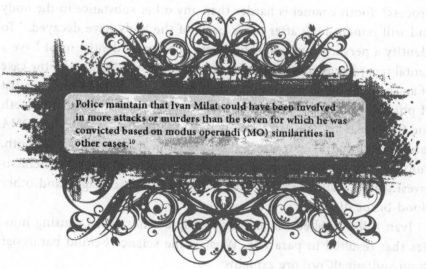

Police maintain that Ivan Milat could have been involved in more attacks or murders than the seven for which he was convicted based on modus operandi (MO) similarities in other cases.[10]

A four-wheel drive vehicle was implicated in the murders. What is the science of identifying tire tracks? According to the Bureau of Criminal Apprehension, "tracks can be collected by photographing, casting, lifting, and/or collecting the clothing from the victim. The tire tracks from the scene can then be compared to tires or known tire impressions from the suspect's vehicle."[11]

One of Milat's victims, Joanne Lees, was able to positively identify her assailant. What is the science of memory? "When we learn something, even as simple as someone's name, we form connections between neurons in the brain. These synapses create new circuits between nerve cells, essentially remapping the brain. The sheer number of possible connections gives the brain unfathomable flexibility. . . ."[12]

The 2001 murder of Peter Falconio was sensationalized in the media in the wake of the Backpacker murders. Given the unusual nature of the attack, and the lack of corroborating evidence (i.e., Falconio's belongings or body), it took the police some days to appreciate the significance of the crime. Survivor Joanne Lees's story was shared as one of survival in a crime of unusual horror against all odds.[13]

The film's release was postponed due to the trial of Bradley John Murdoch, who was convicted and sentenced to life in prison. The real-life murder of English backpacker Peter Falconio served as inspiration for the film. *Wolf Creek* inspired a sequel, *Wolf Creek 2* (2013) and a television series (2016–2017), both starring John Jarratt. Watching these films and shows and reading about the cases certainly has given me pause when thinking about planning a rugged road trip.

A four-wheel drive vehicle was implicated in the murders. What is the science of identifying tire tracks? According to the Bureau of Criminal Apprehension, "tracks can be collected by photographing, casting, lifting, and/or collecting the clothing from the victim, if tire tracks from the scene can then be compared to tires or known tire impressions from the suspect's vehicle."

One of Milat's victims, Joanne Lees, was able to positively identify her assailant. What is the science of memory? "When we learn something, even as simple as someone's name, we form connections between neurons in the brain. These synapses create new circuits between nerve cells, essentially remapping the brain. The sheer number of possible connections gives the brain unfathomable flexibility. . . ."

The film's release was postponed due to the trial of Bradley John Murdoch, who was convicted and sentenced to life in prison. The real-life murder of English backpacker Peter Falconio served as inspiration for the film. Wolf Creek inspired a sequel, Wolf Creek 2 (2013) and a television series (2016–2017), both starring John Jarratt. Watching these films and shows and reading about the cases certainly has given me pause when thinking about planning a rugged road trip.

SECTION THREE
WOMEN KILLERS

CHAPTER SEVEN
Monster (Aileen Wuornos)

Famous novelist and poet Rudyard Kipling once said that the female of our species is more deadly. While statistics would expose this as fantasy, there is something fascinating about women who kill. It flies in the face of what we assume is their natural state: maternal and empathetic. Perhaps they hold a particular interest because, as newswriter Joseph D. Lyons reports, there are so few female killers, especially serial killers:

> According to statistics released by the Department of Justice, since 1980, the number of female murderers has halved, mirroring the decline in male killers. In 2008, only about 1.6 women committed a homicide out of every 100,000 people—not very many at all. In comparison, about 11.3 men per 100,000 carry out a homicide. Black women had the largest reduction in homicide offending rates of all groups measured starting in 1992. The drop was from 11 to 3 per 100,000. Women—in a partial explanation of why women don't commit mass shootings—tend to know their victims, and they tend to murder just one person, not large groups. Only 11.9 percent of female murderers' victims are strangers. Some 41.5 percent of female murderers kill a significant other, be it a spouse, an ex-spouse, or someone they're dating. Another 16.7 percent kill another family member (and 7.5 percent murder a child!). Acquaintances round out the rest of the victims with 29.9 percent. Of all murders with multiple victims, women committed just 6.4 percent.[1]

The world of serial killers, whatever their gender, is a strange and troubling landscape. Male killers like Gary Ridgway and Jack the Ripper, among many others, prey on vulnerable sex workers. But, in a curious reversal in the story of Aileen Wuornos, it is the victim who becomes the predator. After her capture for murdering seven "johns" in a year, Wuornos gave a bleak picture of her life, as cited in Deveryle James's *Zoo of Lust*. "People always look down their noses at hookers. Never give you a chance because they think you took the easy way out, when no one would imagine the willpower it took to do what we do, walking the streets night after night, taking the hits and still getting back up."

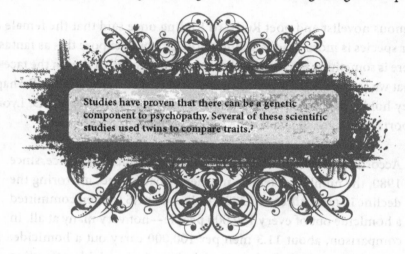

Studies have proven that there can be a genetic component to psychopathy. Several of these scientific studies used twins to compare traits.[2]

Wuornos, like many of her serial killing counterparts, lived through a traumatic childhood marked with sexual and physical abuse. This eventually led to her developing both borderline personality disorder and a psychopathic personality. An assessment tool called the Psychopathy Checklist was used to determine Wuornos's level of psychopathy once she was incarcerated. Developed by Canadian psychologist Robert D. Hare, this checklist is a tool in which to determine if a person has psychopathic traits. It is important to note that

Serial killer Aileen Wuornos.[3]

even if someone scores high on the scale, this does not mean that they will necessarily be a serial killer. Questions on the list include "Do you have a grandiose sense of self-worth? Are you a pathological liar? Do you lack remorse or guilt?"[4] For reference, Ted Bundy scored a 39/40 on the checklist while Wuornos achieved a 32/40. One must score thirty or above to be considered psychopathic. When this diagnostic tool is used by professionals, it can predict if the subject will reoffend, a vital thing to know for parole juries.

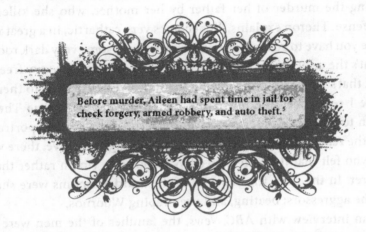

Before murder, Aileen had spent time in jail for check forgery, armed robbery, and auto theft.[5]

In 2003, just a year after Wuornos was executed by lethal injection, a film based on her life came roaring onto the cinematic scene. *Monster*, directed by Patty Jenkins, who would later direct *Wonder Woman* (2017), was noticed by the Hollywood elite and became a box-office smash. This was mostly because of the tour-de-force performance by Charlize Theron who transformed for her role as Aileen Wuornos. She gained thirty pounds, wore fake teeth, and shaved her eyebrows. This change in physicality was an aspect that the media found fascinating, yet it is the emotional work Theron undertook to inhabit Wuornos that is most vital. In an interview, Theron explains her view of Wuornos:

> I see her as an extremely resilient human, probably the most resilient person I've ever encountered in my life. She had to jump back from some really extreme situations in her life. I think we can all relate to the things she goes through in the

film, betrayal and the need to be loved, and the need to fit into a society, and the need to just make life work—that hope of things getting better.[6]

While this may not sound as though Theron is describing a serial killer, she found it necessary to empathize and understand Wuornos in order to become her in the film.

It helped that Theron had lived through her own childhood traumas, including the murder of her father by her mother, who she killed in self-defense. Theron explains, "It becomes very cathartic, in a great way, because you have to go and turn on some lights in some very dark rooms. But that's the only way I know how to work as an actor. It's very freeing. I think that's why a lot of people do therapy. I've never been to therapy because I really feel like my acting is that for me."[7] Charlize Theron went on to win the Academy Award for Best Actress for her portrayal. While the reaction to *Monster* was overwhelmingly positive, there were some who felt that the film showed Wuornos as a victim rather than a murderer. In the film, the majority of the murder victims were shown to be the aggressors; beating, and even raping Wuornos.

In an interview with *ABC News*, the families of the men were less than pleased. "I don't think they cared about the victims' families," said Linda Yates. Her mother was engaged to Gino Antonio, when Wuornos killed him. "[Wuornos] was just a vicious person," said Yates. In fact, the filmmakers didn't talk to any of the victims' families. Mike Humphreys's dad was also murdered. He has a problem with the way the movie was produced: "I don't think that they ought to do this to the victims out there," said Humphreys. "This movie is portraying her as a victim," said Letha Prater. "She isn't. She was not a victim. My brother was a victim."[8] When art and true crime come together, particularly so soon after the events occurred, it can only be natural for there to be disharmony.

Aileen Wuornos shot and killed her victims at close range, ending their lives. The state of Florida ended Wuornos's life by lethal injection in October of 2002. She joined a short list of females executed in the US since the reinstatement of capital punishment in 1976. Only sixteen women have been put to death since then, which represents a mere 1 percent of the over fifteen hundred souls executed. Of these sixteen women, fourteen were killed by lethal injection, including Wuornos. Typically, this injection is a mix of three drugs: a barbiturate, a paralytic, and a potassium solution. The intent is to quickly render the person dead by stopping their heart without causing too much pain. While this may be the aim, many have questioned the efficacy of lethal injection, and most vitally, its morality. Originally, lethal injection was implemented as an alternative to what were considered less humane ways to die, such as firing squad or electrocution. Unfortunately, after several botched executions, lethal injection has been found to be less than perfect. The Death Penalty Information Center, a national non-profit, discusses the current controversy on their website:

> Although the constitutionality of lethal injection has been upheld by the Supreme Court, the specific applications used in states continues to be widely challenged prior to each execution. Because it is increasingly difficult to obtain the drugs used in earlier executions, states have resorted to experimenting with new drugs and drug combinations to carry out executions, resulting in numerous prolonged and painful executions. Even though the issues surrounding lethal injection are far from settled, states are attempting to cut off debate by concealing their execution practices under a veil of secrecy. Recently passed laws bar the public from learning the sources of lethal drugs being used, making it impossible to judge the reliability of the manufacturer or the possible expiration of these drugs.[9]

One controversial drug used in lethal injections is midazolam. While typically used as an anesthesia, as well as for use in people with seizures, it is supposed to help those being executed fall into unconsciousness. The problem, it seems, is that states are giving too much of the drug, and an overdose of midazolam can cause a painful, agonizing death.

Journalist Liliana Segura spoke to Dr. Mark Edgar, a medical examiner for the state of Ohio. Normally, those executed by the state are not sent for autopsy, but after increasing anxiety about the use of midazolam, Edgar wanted answers. Segura writes:

For the past few years, he had been examining the autopsy reports of men executed using midazolam across the country. He found a disturbing pattern. A majority showed signs of pulmonary edema, an accumulation of fluid in the lungs. Several showed bloody froth that oozed from the lungs during the autopsy—evidence that the buildup had been sudden, severe, and harrowing. In a medical context, where a life is to be saved, pulmonary edema is considered an emergency—it feels like drowning. Even if someone is to be deliberately killed by the state, the Constitution is supposed to prohibit cruel and unusual punishment. To Edgar, the autopsies showed the executed men felt the panic and terror of asphyxiation before they died.[10]

Once Dr. Edgar conducted his own autopsy on inmate Robert Van Hook, he saw the carnage for himself. Hook was the twenty-fourth recorded person in Ohio to show signs of pulmonary edema after being lethally injected with midazolam. This also made sense when taking in the accounts of those who had witnessed executions and reported that the men often coughed, choked, and appeared to be in pain.

Armed with this information, Dr. Edgar and his colleagues brought their concerns to the state court. After four days of testimony, Judge Michael Merz found the drug to be inhumane and similar to "waterboarding." A week later, in January 2019, Ohio governor Mike DeWine stayed executions until more research and alternative medicines could be developed.

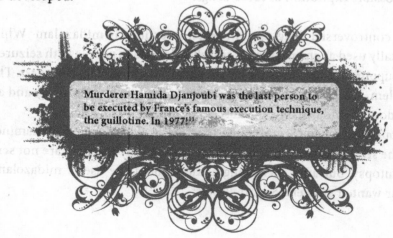

Murderer Hamida Djanjoubi was the last person to be executed by France's famous execution technique, the guillotine. In 1977![11]

This makes us wonder what Aileen Wuornos, and so many people like her, went through in their final moments. *Monster* is a Hollywood film, its creators choosing to empathetically depict Aileen Wuornos. The real truth is much murkier, as she and her seven victims are no longer alive. All that remains is fiction with shades of truth.

CHAPTER EIGHT
Arsenic and Old Lace
(Amy Archer Gilligan)

Long before she was a Marvel superhero, Black Widow was used to describe a woman who lost multiple husbands under mysterious circumstances. There is almost always a financial gain involved, as female serial killers are more likely to kill for money, while their male counterparts are motivated by sex and control. You can read more about this phenomenon in our book *The Science of Monsters* (2019).

The term black widow is derived from the *Latrodectus* genus of spiders, specifically North American black widows (Latrodectus mactans). The females, identified by their dark coloring and hourglass red marking, have the larger venom glands, and can be dangerous to animals, including humans. While it is rare for a black widow bite to be fatal, it can cause distress. According to the National Capital Poison Center, "after a bad bite, severe pain and muscle cramps can start in a couple of hours. Muscle cramps start in the area of the bite (often a hand or foot) and move toward the center of the body. Some black widow bites cause such extreme pain that it's mistaken for appendicitis or a heart attack."[1]

Black widow spiders have a distinctive "comb foot," a series of hairs that are shaped like a comb. This helps them drape silk over their prey before they feast.[2]

The black widow also has a reputation for sexual cannibalism. This is when the female devours the male after mating. Scientists believe this is done to ensure the health of the offspring, though it certainly doesn't happen every time. In fact, male widows like to make certain that their mate has already eaten before they make babies! Who could blame them?

Many female serial killers have proven to be as deadly to their spouses as the black widow spider. Betty Neumar's five husbands all died in bizarre ways. Her third husband shot himself in an apparent suicide—the only problem was that there were two shots to his head.

While the term black widow is synonymous with women murdering multiple lovers, it also refers to women who kill those around them for money. They often prey upon those weaker, like children and the infirm. Stacey Castor killed both of her husbands with antifreeze and attempted to kill her daughter and frame her for their murders. There is also a commonality in their modus operandi, as these women tend to use poison, and will gain finances through either inheritance or by insurance fraud.

Famous black widow Amy Archer-Gilligan, born on Halloween 1873, grew up to be a truly terrifying murderer. She was known to have caused at least five murders, though there is reason to suspect her of many more. A rather harmless looking woman, Archer-Gilligan became a caretaker for the elderly and ill after she married her first

Serial killer Amy Archer-Gilligan.

husband, James Archer. Three years after the couple opened their own rooming house in Windsor, Connecticut, James Archer died of what was described as kidney disease, although Amy had taken out an insurance policy on her husband mere weeks before his untimely death.

Forging ahead with her rooming house, Archer married wealthy Michael W. Gilligan. After only three months of marriage, Gilligan died of what was described on his death certificate as severe indigestion.

Yikes! Yet, he had left his entire estate to his new wife, writing out his four children. This windfall allowed Gilligan to live comfortably, and to continue to invest in her rooming house. Later, it was discovered that Gilligan's will had been forged by Archer-Gilligan to name her as the only beneficiary.

While the husbands of Archer-Gilligan were clearly in trouble, her roomers were also in grave danger. From 1911 to 1916, forty-eight people died under the roof of the Archer Home for the Elderly and Infirm. One suspicious death, that of a healthy man named Franklin R. Andrews, caused neighbors to take notice. Andrews had been seen gardening and attending to yard work as healthy as can be in the morning. That night he died from what was guessed to be stomach ulcers. When Andrews's sister, Nellie Pierce, came in possession of his letters, she realized that he died shortly after giving Archer-Gilligan a large amount of money. This, it seemed, had become a pattern, in which the people staying at the Archer Home for the Elderly and Infirm were found dead soon after handing over sums to their caretaker. Nellie Pierce approached the local district attorney with her suspicions, but law enforcement did not see reason to worry. Angry and desperate, Pierce sought out the media at the *Hartford Courant*. This newspaper was established in 1764 and still runs today as the longest continually running paper in the US. The reporters at the *Courant* were intrigued by Pierce's story and published the first of many articles about Archer-Gilligan and the care home in May 1916. The law finally took notice. CriminalElement.com elaborates:

> Two years after his death, the body of Franklin Andrews was exhumed at the request of the local authorities. The subsequent autopsy of his body found enough arsenic in it to kill several men. The original cause of death, listed as gastric ulcers, was removed from his death certificate and replaced with the ominous "death by poisoning." When other exhumed bodies showed the same type of poisoning, the *Hartford Courant* ran an article about the murders under the headline: "Police Believe Archer Home for Aged a Murder Factory!"[3]

The term "murder factory" stuck, as the bodies of both of Archer-Gilligan's husbands and two other boarders were exhumed. Traces of arsenic or strychnine were found in all. After her murder convictions, Archer-Gilligan was sent to prison, and in 1924 was transferred to live out her days at the Connecticut Hospital for the Insane.

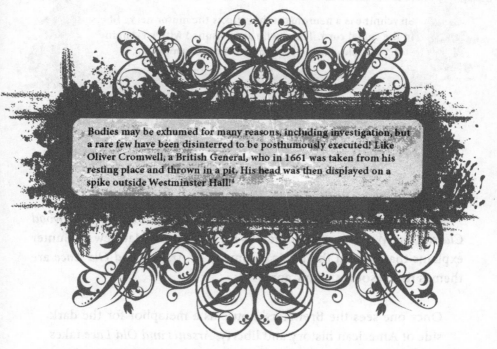

Bodies may be exhumed for many reasons, including investigation, but a rare few have been disinterred to be posthumously executed! Like Oliver Cromwell, a British General, who in 1661 was taken from his resting place and thrown in a pit. His head was then displayed on a spike outside Westminster Hall![4]

On January 10, 1941, the play *Arsenic and Old Lace* premiered on the stage of the Fulton Theatre. Written by American playwright Joseph Kesselring, the play was a dark comedy about two elderly spinsters of the zany Brewster family who killed boarders of their rooming house with poison. Inspired by Amy Archer-Gilligan, Kesselring originally wanted the play to be a drama, yet producers encouraged him to switch to comedy. The play, originally titled *Bodies in Our Cellar*, was the most successful of Kesselring's career, as it was revived several times on Broadway starring the likes of Bela Lugosi and silent film legend Lillian Gish. Most notable is the film adaptation of *Arsenic and Old Lace* (1944) starring Cary Grant and directed by Frank Capra two years before his hit *It's A Wonderful Life* (1946). The film closely follows the play and received positive reviews.

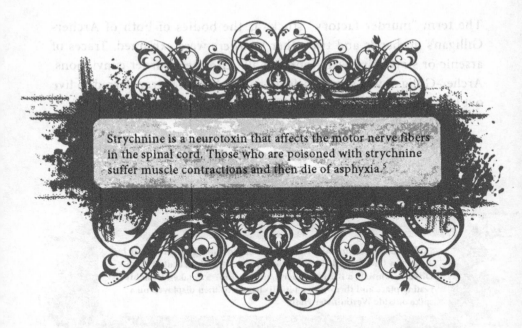

Strychnine is a neurotoxin that affects the motor nerve fibers in the spinal cord. Those who are poisoned with strychnine suffer muscle contractions and then die of asphyxia.[5]

In his book, *The Capra Touch: A Study of the Director's Hollywood Classics and War Documentaries, 1934–1945*, author Matthew C. Gunter explains that behind the farcical comedy of *Arsenic and Old Lace* are themes that resonate with Capra:

> Once one sees the Brewsters' house as a metaphor for the dark side of American history and liberty, *Arsenic and Old Lace* takes on a much greater relevance. Capra, either subconsciously or consciously (my guess is the latter), understood that Kesselring's play is really about the nature of freedom in America. The contrast of the sweet disposition of Mortimer's aunts with their horrific actions illustrates the problem of America itself. While America claims to be an optimistic beacon of liberty to the entire world, it can also be a place of extreme violence and prejudice."[6]

While Capra's play may have been changed significantly to center on the silly Brewster aunts for comedic effect, at the heart of *Arsenic and Old Lace* is what happened in a Connecticut boarding house. As Gunter expressed, there is something wholly disparate between the edifice of the outside of both the house and the woman, while murderous evil boils underneath.

As we explore the intersection of art and science in true crime, it is important for us to reach out to artists who are invested in telling victim's stories. Olivia Gatwood is a nationally recognized poet, artist, and educator in sexual assault prevention and recovery. Her poetry collection *Life of the Party* (2019) is a chilling and beautifully written look at the reality of murder from multiple perspectives. We were able to pick Olivia Gatwood's brain about the varying true crime topics in her book. It is important to note that when referring to men versus women in this interview, and in this book, it is to amplify the perceived societal differences between the two.

Meg: "In the author's note of your poetry collection *Life of the Party* you discuss both your fascination with consuming true crime, as well as frightening, criminal incidents that happened to you and your friends. Why do you believe women are drawn to true crime even though we are, overwhelmingly, the victims?"

Olivia Gatwood: "I can only pathologize, but the most accurate understanding I've come to is that it's a sort of twisted form of representation. There's something cathartic about seeing people who look like you experience something that you've been afraid of your whole life, especially if that fear is something other people have told you is irrational. There's also a way in which I think a lot of women consume true crime as a kind of preparation, which is complex because that implies that there is a right or wrong way to exist, or a right or wrong way to avoid getting murdered, which isn't true. But I do think we put ourselves in these stories when we hear them, we take note of what happened, and we operate accordingly with that new knowledge."

Kelly: "Poetry and spoken word have been your ways of expressing yourself. What sort of emotions about crime against women have you processed through your art form? Have you found it to be therapeutic?"

Olivia Gatwood: "*Life of the Party* was my way of understanding my own fear, both by critiquing it on a societal level and a personal one. It was deeply healing to write that book because it helped me process so many things that were muddy in my brain. Finally, I was able to put them into the world in a cohesive and polished way."

Meg: "Often, true crime centers on the male murderer, sometimes even placing that serial killer on a pedestal, making them a sort of icon. How can we fight against this? Is it about focusing on the victims?"

Olivia Gatwood: "We need to focus on the victims, and we need to understand that violence against women is not sensational or rare. When we talk about Ted Bundy like a celebrity, we are also talking about a serial rapist and murderer. In most cases of homicide against women, it is a person that the woman knew, often a boyfriend or husband. And so, it's not just a stranger in a mask who will later become famous for the crimes he committed. It's often much more personal than that. We need to remind ourselves of this if we actually want to combat a culture that romanticizes this violence and the people who commit it."

Kelly: "Can you describe your interest in Aileen Wuornos? What was it about her story that led you to write several poems about her?"

Olivia Gatwood: "Aileen occupies every role that *Life of the Party* seeks to interrogate. She was both the murderer and the murdered. She killed several men and was then killed by the state. She was a victim of abuse, she was queer, she was a sex worker. I don't want to excuse the harm that Aileen caused, but I do want to approach it with complexity, because I feel empathy when I read about her. To me, she was a woman who was driven to violence by violence. And that is far more nuanced than she's often given credit for."

Kelly: "Your poem about JonBenet Ramsey really hit me. It is an indictment about how the media, and the world, treat certain victims (white, cute, blonde) so very differently than others. Do you think we've grown as a society since Ramsey's death? Or do you think there is still a long way to go in the media and beyond?"

Olivia Gatwood: "I think our discourse and understanding around race have evolved in certain ways. It's more common knowledge that women of color, specifically Indigenous and Black women, even more specifically Trans Black women, are most likely to be victims of homicide. But in terms of the media and the consumption of true crime, I do think the obsession with the white, cis, blonde victim is still very prevalent. There's a long way to go because, ultimately, the genre of true crime is still a genre of entertainment media—it's still about beauty, sex, views, ratings, demand. Until we confront that, we won't see a shift in the stories it chooses to highlight."

Meg: "'When They Find Him'' is a poem about the Golden State Killer. But it's really about how men and women have such vastly different life experiences. How women often see the worst in their male partners and family members, while the males in their lives are blind to their fellow males' aggressions. Do you believe that until men understand and empathize with our point of view, there will be no change?"

Olivia Gatwood: "I don't think we need to solely rely on men for there to be change, but I do think men's empathy is a crucial element in creating a safer world for everyone else."

Kelly: "Can you share your thoughts on male killers versus female killers? There are studies that suggest female serial killers kill for different reasons than their male counterparts. What do you think about the motives of female killers? Do you think their motives are less sinister?"

Olivia Gatwood: "I don't have the most well-researched viewpoint here, but from what I understand and have seen, it seems like most women who kill are driven to that point after surviving abuse. It doesn't seem as common for women to commit murder out of perverse fantasy. Again, that's not to justify it, but to complicate it. Of course, there are exceptions. But I think it's telling that so often the shows about women who kill have titles like *Snapped* (2004–), which imply that the killer had lost patience with something, that something pushed her."

Meg: "Please share with us your new and upcoming projects!"
Olivia Gatwood: "My novel, *Whoever You Are, Honey*, will be out in 2022. It's a feminist psychological thriller about beauty-induced hysteria."

Thank you to Olivia Gatwood for this insightful interview!

CHAPTER NINE
America's First Female Serial Killer (Jane Toppan)

There is no way to tell, when a baby is born, if they will grow to become a monster. The child seems like a blank slate on which their parents, family, and peers can write. Yet, the construction of a personality, particularly that of someone as complicated and nefarious as Jane Toppan, must be an amalgam of both "nature" and "nurture." This debate of whether we are hardwired to be who we are, or if our environment creates us, has raged on in psychological circles. Those who believe that our behavior and personality are formed almost entirely by genetics are called nativists. *Simply Psychology* explains:

> Examples of extreme nature positions in psychology include Noam Chomsky, who proposed language is gained through the use of an innate language acquisition device. Another example of nature is Freud's theory of aggression as being an innate drive (called Thanatos). Characteristics and differences that are not observable at birth, but which emerge later in life, are regarded as the product of maturation. That is to say, we all have an inner "biological clock" which switches on (or off) types of behavior in a pre-programmed way. The classic example of the way this affects our physical development are the bodily changes that occur in early adolescence at puberty. However, nativists also argue that maturation governs the emergence of attachment in infancy, language acquisition, and even cognitive development as a whole.[1]

Greek philosopher, Galen, believed that a human's traits depended on how much of four liquids, or humors, filled their body. These were blood, phlegm, yellow bile, and black bile.[2]

On the other end of the nature versus nurture spectrum are the empiricists. They believe that we are formed by our experiences and how we were raised. For instance, in Albert Bandura's 1977 study, he posits that we learn aggression from those around us. It may be from parents, peers, or even from what we see on television. In Bandura's "Bobo doll experiment" he observed children's aggression in a nursery or playroom environment. He then placed them in control groups, and had them watch as different people (adults and children) hit, yelled, and abused the Bobo doll. When the kids were left with the Bobo doll, they recreated what they had witnessed, to varying degrees. For example, boys were more likely to imitate the aggression of men than women. The study concluded with the evidence that there is a social component to how we behave. It is probably more likely that we fall somewhere in the middle of the nativists and empiricists. Although it is impossible to say how much of each contributes to our being.

Nature versus nurture becomes even more fascinating when we examine people who exhibit far outside what is considered normal behavior. Jane Toppan is a prime example of this. Born in 1854 as Honora Kelley, Toppan's childhood was particularly tragic. Her parents were Irish immigrants at a time when the Irish were considered undesirable. After her mother's death, Toppan's father went insane. Rumor has it that he sewed his own eyelids shut after dropping Toppan and her sister at a home for the insane and forgotten. At age eight, Toppan was taken to live at the Toppan residence, where she took their name, though instead of being cared for as a child, she was treated as an indentured servant in the vein of Cinderella. As told in *America's First Female Serial Killer: Jane Toppan and the Making of a Monster* (2020), Toppan's life continued to unravel. It wasn't until she became a nurse at Cambridge Hospital that she took control of her own fate. Through the warped prism of her mind, nurse

Serial killer Jane Toppan.

Toppan abused that control by playing with lethal medicines, beginning her killing spree on the vulnerable patients who needed her most. A nativist would argue that her father's mental illness was the clear cause of her murderous ways, while an empiricist would say it was the severe neglect and abuse at the hands of her adopted mother that led Jane Toppan to kill.

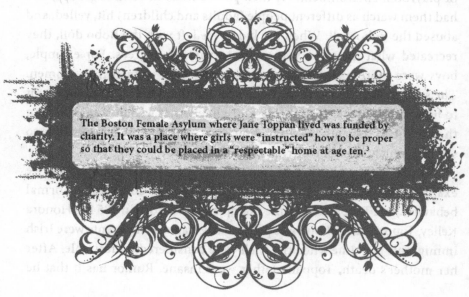

The Boston Female Asylum where Jane Toppan lived was funded by charity. It was a place where girls were "instructed" how to be proper so that they could be placed in a "respectable" home at age ten.[3]

The dynamic of nurses killing their patients is unfortunately a well-documented phenomenon. Although rare, these serial killers find a fertile killing ground at the hospital in which they work. They are depended upon by people in often the most trying times of their lives, expected to soothe, aid, and heal. The dichotomy of what these murderous nurses portray to the world, versus their true nature, is frightening to even the most hardened cynic.

Raised in Fall River, Massachusetts, not far from Cambridge, and the same town where Lizzie Borden reigned, Kristen Gilbert became a serial murderer just like Jane Toppan. Nearly one hundred years later, Nurse Gilbert killed four men (and perhaps more) at the Veterans Affairs Medical Center by injecting lethal doses of epinephrine into their IV bags. Epinephrine is a synthetic drug that acts similarly to our natural adrenaline. It can be used to revive a dying patient in cases of cardiac arrest, but can kill a patient if used irresponsibly. Gilbert's motive in giving patients this drug was so that she could be seen as a hero. She would inject patients with a lethal dose, bringing them to near death, just before swooping in and saving them. She received much fanfare, something she was clearly seeking, but her four known victims weren't so lucky to be brought back from the brink. Once her fellow nurses became concerned

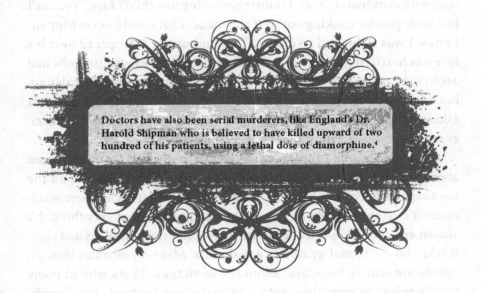

Doctors have also been serial murderers, like England's Dr. Harold Shipman who is believed to have killed upward of two hundred of his patients, using a lethal dose of diamorphine.[4]

about the growing number of deaths and missing epinephrine on their ward, Gilbert could no longer hide from law enforcement. Gilbert now resides in a Federal Medical Center in Fort Worth, Texas, after being convicted of four life sentences.

After killing an undetermined number of hospital patients, Nurse Jane Toppan began killing neighbors and close friends. Her harrowing story is written in vivid and compelling detail in *America's First Female Serial Killer: Jane Toppan and the Making of a Monster*. We had the great fortune to speak with the book's author, Mary Kay McBrayer.

Meg: "In the introduction to *America's First Female Serial Killer*, you describe how you came to know the history of Jane Toppan. You said 'the more facts I knew about Jane, the less I knew her.' Could you elaborate on this statement? Why do you think it was difficult to get to know her? And how did you prepare to write from her perspective?
Mary Kay McBrayer: "Fresh out of college, I served with a volunteer group at a residential mental health facility for emotionally disturbed children, and even though that term of service was brief, it was the most rewarding (and least sustainable) period of my life. I was assigned a group of fourteen- to eighteen-year-old boys, most of whom were wards of the state with criminal records. I remember saying to a child's face, 'You can't just apologize for choking out your classmate. That should never happen.' I knew I was right, and so did he. The vindication that spread over his face was horrifying, the way he calmly explained how many times he had asked to be taken into seclusion because he could feel himself escalating, but our resources were such that we couldn't accommodate what he was asking. It was like he was waiting for confirmation that he was a monster, even though his circumstances pushed him into it.

Criminal records or instances of bad behavior like the one I stated above are all that most people saw in those boys, and that was all the boys expected people to see. The saddest aspect of that cycle is how much easier it is to see only the crime. I watched myself do that very thing. It's almost an act of self-preservation when people write kids off as bad eggs. It takes no emotional gymnastics to do that. Make no mistake, though, crimes *are* crimes, inexcusable and full of victims. That's why so many people refuse to empathize with criminals. I understand that thought process firsthand, and it is a comfortable position, that irate and righteous moral high ground. Looking only at the newsworthy facts of someone's life, though, tends to cut off potential for growth or improvement at any level. Like I said, it's easier.

What I saw happening in my own experience of adjudicated children is also what I experienced with Jane. It's very difficult to come to terms with conflicting truths: this kid tried to physically hurt someone who had done nothing to him, but this kid had been physically hurt by someone to whom he had done nothing. It's almost like—and this exercise carries over into my experience of Jane's story, too—I had to zoom out of the situation to say, both of these things are true, and they can't coexist, but they do. It was hard to get to know her, frankly, because, well . . . someone like that? You don't want to get too close to them."

Kelly: "Jane's childhood was tragic, to say the least. You said that 'the human was driven out of Jane.' Do you believe female and male killers are formed by similar struggles? Or are there gender-specific differences in the formation from child to 'monster?'"

Mary Kay McBrayer: "Boys and girls both have a ton of societal pressure (especially from conservative families) to fit a mold, which no one can do. It should be noted as well that in our present culture, 'boy' and 'girl' are the two options provided, though we know that those are not the only two genders. Those standards or 'norms' are much more to blame for violent behaviors than biological sexual difference.

Though, in an overgeneralized, culturally acceptable heteronormative binary, men have historically experienced a very set standard of 'manliness,' which makes for a whole pile of issues, of course. In my professional experience, that cultural expectation leads to the idea that when boys have an issue, they should fight, physically. And then, usually, it's over. Win, lose, or draw, that's usually the end. In my personal and professional experience of the second side of that binary, girls have a longer memory. Culturally, we are not allowed to fight. Rather than take a swing at her jaw, as a girl, I should be more likely to slip her weight-gain bars like in the film *Mean Girls* (2004). Forever. Genders don't always fit into those two roles, but the fact that those roles exist creates problems for almost everyone. Women still kill, but are more likely to use nonviolent means. Women tend to choose an oven or pills instead of a gun, for example.

That said, I don't think criminal behavior is biological or attached to one's sex. It is learned. The escalation into criminal behavior, from child to 'monster,' like you said, seems more related to an authority figure making you hate yourself, or parts of your person. In Jane's case, she was Irish in America at the turn of the century. She was olive-skinned and smart and fun, and that was just outrageous for a person of her class. So, they changed her name, trained her out of her accent, kept her in solitude, and never let her rise above her station. There were severe consequences when she overstepped that role. I don't mean to overgeneralize, but the cliché 'hurt people hurt people' is accurate. The catch is, not *all* hurt people hurt people, and that's something worth investigating."

Meg: "How did you research the scientific aspects of your book?"
Mary Kay McBrayer: "Fortunately, I had access to a university library, and I was able to get so much of the contemporary source material through interlibrary loan. Plus (did you know this?), if you just go to Emory University's library and ask to see something specific, they'll let you in. I did that with the microfilm. Actually, the newspapers you see on the cover of *America's First Female Serial Killer: Jane Toppan and the Making of a Monster* I found during those searches. Same with the ones in the front matter, and the ones transcribed verbatim. The ink splotches in the design are to lend authenticity where I couldn't make out the words in the original documents.

The editorializing of events was something I had to wade through as well. I read everything I could find about the era's contemporary medical science, and a lot of it I found through bibliography hopping. For example, when Harold Schechter said in *Fatal* (2003) that Amelia Phinney had an ulcer burned off her uterus with silver nitrate, I flipped to the back of the book for Schechter's source material and then located and read that. By the way, I was beyond shocked to learn that until recently, for tampons and sanitary pads, women just had to use whatever was around because polite society didn't really acknowledge menstruation. Can you imagine? And—oh my God—I was just searching images of turn-of-the-century pill bottles on Pinterest one day, to grasp the aesthetic of what they had looked like before plastics, when I saw those chocolate-covered strychnine tablets.

A lot of my research was serendipitous like that. When I was looking for Jane's confession in the paper, it'd be run alongside an ad for over-the-counter-arsenic or something wild, which I then verified with mouth ajar. Pretty much all the medical science from the Gilded Age is horrific. I mean, they barely knew how to wash their hands. Let alone medicine toward women. Ugh, sepsis was incredibly common because their sanitation methods were so haphazard. When all else failed, though, or I just couldn't trust my own eyes, I'd run it by my best friend Mimi. She is a nurse (at Grady Hospital in the ICU), and at the time, she lived next door. So, I could just be like, 'Hey girl, you want some breakfast? Also, do y'all still feed people charcoal to make them purge? And is it like, a briquette, or . . . ?'"

Kelly: "Do you think the media portrays female serial killers with more empathy? Do you think Jane Toppan deserves more understanding and empathy than a serial killer like Ted Bundy?"

Mary Kay McBrayer: "This is a truly great question. I'm not sure either of them deserves empathy, to be honest, but they both certainly deserve to be studied, and the world deserves to learn about them if only to get better at early intervention. Ultimately, I don't think our culture is empathizing with Bundy-like killers so much as sensationalizing and almost idealizing them, and that is dangerous. I really do not understand this ongoing cultural fascination with Ted Bundy. I go to a dance studio regularly, and I can't even list the number of younger clients there who (when they heard about this book) told me how attractive Ted Bundy was. I was so bewildered . . . like, okay, number one, Zac Efron does have a huge range and he is very watchable, so I understand being fascinated with his performance in the dramatized version, *Extremely Wicked, Shockingly Evil and Vile* (2019). But the real Ted Bundy? Girl. Where to begin? Where to end?"

Meg: "I was chilled by the account of Jane's attempted murder of Amelia Phinney in your book. Can you explain why you think Jane killed in the manner she did? It seems paradoxical that she was a poisoner yet liked to be so physically close to the dying."

Mary Kay McBrayer: "Amelia Phinney's testimony is one that helped truly convict Jane, because Amelia Phinney survived. (If you want to read the most factual account of that, you can check out Harold Schechter's *Fatal.*) As if it wasn't enough to have her uterus burned with silver nitrate, while she was recovering she was sexually molested by her nurse, who was trying to murder her, and then she had to testify against her years later in a pristine Victorian court. What a champion.

Clearly, I want to identify the most with Mrs. Phinney because she is the truest version of a hero, but the nature of the book took me in the opposite direction. It seems as though, at that stage of escalation, Jane was less interested in taking life and more enthralled with controlling life. She enjoyed the feeling of inflicting death and then reviving her victim just before it was too late. Jane was very smart. It seems like this kind of depraved curiosity happened in response to her inability to put her skills to good use—not only because of that, but in correlation to it."

Kelly: "Many serial killers, male and female, have numerous movies devoted to them. Why do you think the story of Jane Toppan hasn't resonated in the film world? Would you like to see it become a film?"

Mary Kay McBrayer: "Oh. Yes. I would love to see this in film form. I wonder if the reason why Jane Toppan hasn't resonated on screen yet is because stories like this were until recently approached like this: 'Okay, that happened. What do we do with that?' It's not a feel-good story, there is no true moral takeaway, and the sexuality of the murders is also uncomfortable. Don't get me wrong, I would watch the hell out of a movie like that, but I wonder if a story like that one is just more of a gamble when production companies can, instead, do another install-ment of whatever trending superhero and definitely hit that return on investment, you know?"

Meg: "I loved your book. It feels as if you achieved bringing both Jane and her victims to vivid life. Tell us how you expertly wove deduction with real facts to tell a compelling story. How do you balance both?"

Mary Kay McBrayer: "Oh, thank you so much for saying that, Meg! I'm so glad you enjoyed it. Because I studied creative writing and (auto) biography theory, I knew going into it that when writing creative nonfiction, no one can ever make everyone happy. I knew going in that the story-making part of this writing process would upset some readers, whether it was in the recreation of dialogue or just having to decide things based on deduction. That is the nature of the beast. It's important to me, though, to always establish a contract with the reader. I would never intentionally mislead or misrepresent a situation—to do so renders prose fictional, that's where the line is in my view—but I think it's important to acknowledge, too, that no situation is one-sided, and when there is more than one perspective involved, there is no objective truth of a scene. Everything goes through a perspective, and that paradox is troubling, like I mentioned before. Not everyone agrees with that balance . . . or even wants to acknowledge that the balancing act exists. Long answer shorter: I searched everywhere I could for the facts. When I couldn't find what I needed, I researched around it and decided on the most likely instance . . . or, honestly, the one that felt the truest."

Kelly: "Tell us about your current and future projects!"

Mary Kay McBrayer: "My second book is in the works—it's about Madame Stephanie St. Clair, the 'Numbers Queen' of Prohibition-era Harlem. Not many people have heard of her, but they know of Bumpy Johnson (her enforcer), Frank Lucas (his protégé), and Dutch Schultz (the white gangster who tried to take over her territory and failed), so stick with me for more on that. I also cohost a comedy podcast that analyzes horror movies called *Everything Trying to Kill You.* You can follow me on Twitter and Instagram, too."

Thank you to Mary Kay McBrayer for this incredible insight into Jane Toppan. We can't wait to read her new work!

SECTION FOUR
DOUBLE LIVES

CHAPTER TEN
The Stepfather (John List)

For one to be considered a serial killer, they technically have to have killed three people in more than one incident. Therefore, the type of murderers known as "family annihilators" are a creature all their own. While men like Joseph DeAngelo stalked the streets of Sacramento in order to gain entry into strangers' homes, a family annihilator is arguably more terrifying. These predators lay in wait like a crocodile beneath the water, surfacing only when it is time to show their true nature: a razor-sharp bite that kills. Worse, family annihilators like John List don't take out their rage on strangers, they wipe their entire families from existence.

Family annihilators have been studied to discover similarities. They are mostly male, in their thirties, and usually have no known mental problems or trouble with the police. One British study found that these killers could be placed in four distinct categories:

> Self-righteous killers hold the mother responsible for the break-down of the family and will often call her before to explain what he is about to do. Disappointed killers believe their family has let them down, and the killing could be sparked by something like children not choosing to follow religious customs. Anomic killers see the family as a symbol of their own economic success, but if they suffer some kind of economic failure—bankruptcy, for example—the family no longer serves this function. Paranoid killers are often motivated by a desire to protect their family from a perceived threat, such as having children taken away by social services.[1]

One recent example of a family annihilator is Chris Watts. In 2018, Watts killed his pregnant wife and two young daughters. The media was fascinated with the Watts family case, as Shanann and her daughters were such inno-cent and empathetic victims. Before Watts confessed to the murders, he told a chilling lie: Shanann had killed the two girls, and he had killed her in a fit of rage. This, of course, was not true, and has a ring of self-righteousness to it. Watts will spend the rest of his life in prison, yet his punishment will never rectify what he has done. Many have been left confused by the Watts murders, as there just seemed to be no reason or motive.

The first time I (Meg) was introduced to John List was through the television movie *Judgement Day: The John List Story* (1993). List was played (ironically) by Robert Blake, who would later go on to be convicted of the murder of his girlfriend, Bonnie Lee Bakley. In this film, I was introduced to the idea that a father could kill his children. While I was used to fearing strangers and the boogeyman, this particular horror had never occurred to me. Naturally, the concept unnerved my young brain, as I pondered if any man in my life could become so cruel. I believe it was this film that not only sparked my fascination with true crime, but also inspired my love of *The Shining* (1980), in which a supernatural element adds to a father's murderous impulses.

To understand the films that were inspired by John List, we must first meet the man. In 1971, List and his wife, Helen, had been married for twenty years and he was a father of three. He lived in an impressive mansion in Jersey City, New Jersey, where his elderly mother resided in the attic rooms. His balding head and glasses made him appear as innocuous as any middle-aged man. Like many family annihilators, List had no history of mental health problems or brushes with the law. In fact, he was a veteran of the Korean War. On November 9, 1971, John List systematically killed all five members of his family.

After his three teens left for school, List shot both his wife, Helen, and mother, Alma, in the backs of their heads. He then waited for daughter Patricia, sixteen, and son Frederick, thirteen, to return home. He shot them both. After watching his fifteen-year-old's soccer game, all the while knowing that the rest of his family was dead, he took youngest son John Jr. home and killed him last. It was clear that List had planned these murders, as he had called the school in advance to let them know they were going on a family trip. He had also drained his bank account and prepared for his escape.

One aspect of the List murders that haunted me, as it was depicted in *Judgement Day: The John List Story*, is that List wrapped each of his family members and placed them in the ballroom of their Victorian home. He then turned off the heat in order to slow the bodies' decay. Next, before he disappeared into the night, John List left the radio on a classical music channel, filling the house with eerie sound.

It was not until December 7 that the local police were finally urged to enter and check in on the List family. After their horrific find, the manhunt was on for John List. He had ruined every picture of his face in the home, which at a time before social media, made it almost impossible to achieve a perfect sketch of the fugitive. Despite the police's dogged pursuit of List, the trail was cold. Years passed and there was no clue where John List had disappeared to. It was as if he had simply slipped away into the shadows.

It was in the spring of 1989, eighteen years after the murders, that *America's Most Wanted* (1988–2012) devoted an episode to John List. After recounting the horrific murders that happened in 1971, host John Walsh presented a clay bust that had been sculpted by forensic artist, Frank Bender. By using both science and art, Bender created the likeness of List to show how he might look after aging nearly twenty years.

Frank Bender began his career in a rather peculiar manner, studying corpses in the Philadelphia morgue to improve his sculpting skills. When an unidentified woman killed out-side the airport was brought into the morgue, Bender used his clay to reconstruct her dam-aged face. His bust helped her family recognize her and helped the police catch her murderer. This success led him to create life-size bronze statues for New York's African Burial Ground National Monument, using skulls found on the site to create proper likenesses.

Sculptural portrait heads from classical antiquity are sometimes displayed as busts. However, these are often fragments from full-body statues, or were created to be inserted into an existing body, a common Roman practice.[2]

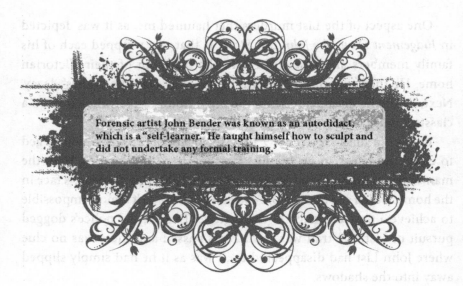

Forensic artist John Bender was known as an autodidact, which is a "self-learner." He taught himself how to sculpt and did not undertake any formal training.[3]

Frank Bender's business manager described one disturbing but necessary procedure he used to create. "He would get the skull, sometimes boiling away the remaining flesh, and then mold a face. The process took about a month, but much of the work was done after a flash of inspiration."[4] Eventually, Bender's impressive skill led him to assist in identifying remains and creating busts of fugitives like John List.

For Bender's last project before his death from cancer a month later, he assisted in identifying a woman found murdered ten years earlier in the Pennsylvania woods. In a 2011 article for *The New York Times*, journalist David Stout explains the balance of art and science that Bender so deftly walks:

> When Mr. Bender measures a forehead and the distance from, say, eye socket to nostril hole, he starts to see a face. From statistics and experience, he surmises how thick the tissue must have been, the shape of the nose, the fullness of the lips. But there is more to it, as this case demonstrates. Perhaps, a visitor suggested, the woman was a drug addict or prostitute who had dropped out of conventional society. The fact that she has been unidentified all these years pointed to such a background. Right? Wrong, Mr. Bender said. The extensive dental work, including a root canal

and crowns, suggested that she'd had resources, sophistication, self-esteem. Maybe, he said, 'she got a divorce, was feeling her oats, wanted to start a new life—and met the wrong guy.'[5]

While art is at the center of Frank Bender's creations, science is clearly an integral aspect of his work as well. It is this blend of art, science, and even psychology, that made Bender so successful. He took tender care to come to know and understand the victim.

In the case of fugitive and murderer John List, Frank Bender had to work on the assumption that he was alive and that he, naturally, had aged. Illustrating age progression today is almost exclusively done through digital enhancement of photographs, but Bender was notoriously unimpressed by computers. Even in 2011, he felt as if they could not capture the humanity of the victims, or even the perpetrators.

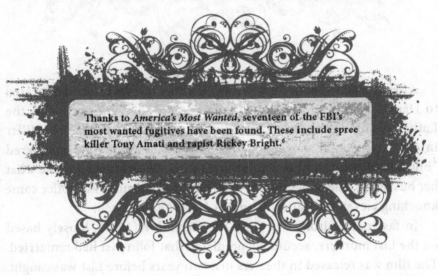

Thanks to *America's Most Wanted*, seventeen of the FBI's most wanted fugitives have been found. These include spree killer Tony Amati and rapist Rickey Bright.[6]

In 1989, at the behest of *America's Most Wanted*, Bender began his bust of John List. He used the only blurry photographs of List available, as well as witness accounts of the man himself. The age progression had to be specific to List. What sort of lifestyle did he live? Would he be overweight? Bald? By using his knowledge of sociology and genetics, Frank Bender created a bust that was remarkable in its likeness. The bust was such a success that it was less than two weeks later when John List was arrested in Virginia. Thanks to List's neighbor, who immediately noticed the resemblance of Bender's bust, John List finally faced justice.

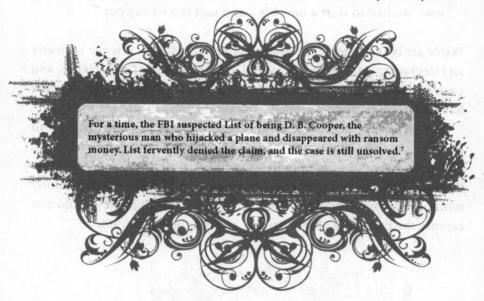

For a time, the FBI suspected List of being D. B. Cooper, the mysterious man who hijacked a plane and disappeared with ransom money. List fervently denied the claim, and the case is still unsolved.[7]

After the cold-blooded murder of his entire family, List had relocated to Denver where he lived under the name Bob Clark. Active in the Lutheran church, he worked in accounting, and married Delores Miller in 1985. The couple moved to Virginia in 1988, where mild-mannered "Bob Clark" took another job in accounting. Delores had no clue what her husband's true name was, or what he had done, until the police came knocking eleven days after *America's Most Wanted* aired.

In fascinating irony, the film *The Stepfather* (1987), loosely based on the List murders, accurately predicted that John List had remarried. The film was released in theaters just two years before List was caught. It is the chilling tale of a family annihilator with many aliases played by Terry O'Quinn.

The film begins with a bloody scene. The Stepfather is calmly bathing and dressing after the brutal murder of his family. He simply walks away from the carnage, just like John List, and disappears. He can hide in plain sight because he doesn't look like the boogeyman. He has the charm of a family man, a sort of average face and mild personality that no one would suspect. Soon, the Stepfather has wormed his way into another family. He has moved in with Stephanie (Jill Schoelen) and her mother, Susan (Shelley Hack). While Stephanie is in love with her new husband, Susan becomes suspicious of this stranger in her home. As Susan's suspicions reveal themselves to be true, it becomes clear that she and her mother are in true danger.

For me, the most frightening scene in the film is when the Stepfather, looking at his reflection in a mirror, furious and with murder in his eyes, asks, "Who am I here?" It echoes the life of John List, who tried to shed the life of his past to become Bob Clark. Yet, both the fictional Stepfather (who returns in sequels), and John List, could not escape the truth of what they had done.

The law finally came to seek justice, and John List lived out his remaining years in prison. Before his death in 2008, List shared his warped reasoning for killing his family. Because of financial trouble, he believed his only options were to either accept welfare or send his family to heaven. Welfare was, to List, an embarrassment. He reasoned that his family was better off dead than criticized by their neighbors. When asked why he didn't take his own life, John List explained that suicide would bar him from heaven. He wanted to make sure that he would see his family in the afterlife.

Obviously, there is no proven logic behind John List's bizarre explanations. Like the senseless murders of the Watts family, there is simply no motive that makes sense. Though, we can all agree that the idea of a murderer hiding in plain sight, perhaps in our very own home—even sitting across from us at the dinner table—is more terrifying than any stranger in the shadows.

CHAPTER ELEVEN
I'll Be Gone in the Dark (The Golden State Killer)

"The doorbell rings.

No side gates are left open. You're long past leaping over a fence. Take one of your hyper, gulping breaths. Clench your teeth. Inch timidly toward the insistent bell.

This is how it ends for you."[1]

The final words of Michelle McNamara's book *I'll Be Gone in the Dark: One Woman's Obsessive Search for the Golden State Killer* (2018) take on particular poignancy because that imagined moment of his apprehension did, indeed, come for Joseph DeAngelo. Yet, McNamara did not live to see it. Nearly two years before the publication of her research, at only forty-six, Michelle McNamara died in an accidental overdose. Within the pages of *I'll Be Gone in the Dark*, the reader comes to know McNamara and her fixation on the Golden State Killer. In fact, she coined that particular nickname, as the unknown assailant was previously known as the East Area Rapist or the Original Nightstalker. Because of the memoir aspect of her book, we come to understand that McNamara's interest in violent crime was sparked by a murder in her Chicago neighborhood. This watershed moment in her life led to the creation of her popular crime blog, *True Crime Diary*, in 2006. She was eventually drawn to a particular set of unsolved rapes and murders on the West Coast.

From 1973 to 1986, one man was responsible for wide-scale horror in the Bay Area of California. His brutality escalated, burgeoning from burglary, to rape, and then to murder. It was a frightening time for those who lived through the madness. Small communities went on alert, as this stalker would slip into any random window and cause irrevocable harm.

One of the first victims, Jane Carson Sandler, discussed her trauma with the *Los Angeles Times*:

> Certain things always trigger flashbacks to that night in 1976 when DeAngelo confronted her with a butcher knife as she snuggled in bed with her three-year-old son after her husband left for work at a nearby military base. She can't go skiing, for fear she'll see someone in a ski mask like the one DeAngelo wore. The sound of a helicopter is another trigger, because "after the attack the helicopters would fly over every night with spotlights on the ground, looking for DeAngelo."[2]

At first, the Golden State Killer targeted women alone, or with small children, like Sandler. As he grew more confident, he found sick enjoyment in raping women near their husbands. He would achieve this by tying up the man and sometimes placing dishes on his back. If the husband moved, causing the dishes to fall, the Golden State Killer would assure him he, and his wife, would die. This threat became real in several instances, like for Keith and Patrice Harrington in 1980, and for Cheri Domingo and Gregory Sanchez in 1981. These untimely deaths left their families and community reeling. And for decades, after dozens of rapes and thirteen murders, their perpetrator walked free.

While Michelle McNamara worked every angle that she could to find the man responsible, California law enforcement worked, too. Without a badge, McNamara focused on interviews and research. She studied maps, read thousands of pages of police reports, and even searched eBay for possible stolen goods from Golden State Killer victims' property.

Law enforcement held vital evidence in the discovery of the killer's identity. In the 1970s and eighties, the Golden State Killer could not have predicted that leaving his semen and saliva at the scenes would lead to his capture. That is the glory of science! It took advances, and a lot of grieved waiting for the victims and their families, but DNA finally unmasked the killer. Without a DNA profile match in the national CODIS database, those working to capture the Golden State Killer had to find another avenue to success. In a perfect harmony of science and police work, the road led to Joseph DeAngelo; father, grandfather, and former police officer. Interestingly, McNamara had theorized that the Golden State Killer may have been in law enforcement.

So, how did they trace the DNA to DeAngelo?

They worked backward. Scientist Barbara Rae-Venter input the DNA into a database, finding people who were related to the assailant, even distantly. By working through the family tree, and then by using criminal profiling to outline probable age, etc., Joseph DeAngelo's name arose. Confident that he was a decent suspect, officers were dispatched to wait outside DeAngelo's home, follow him until he left a DNA sample, like his spit on a discarded straw or soda can, and run the DNA for a match. DeAngelo lived a rather hermit-like lifestyle, so apparently it took a while for a proper item. It's important to note, there were other suspects pursued through familial DNA, that led to a non-match.

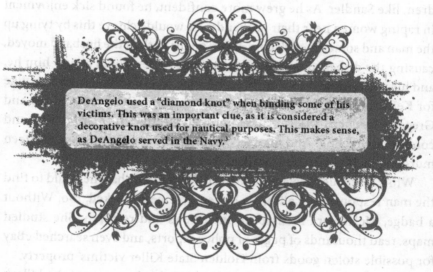

DeAngelo used a "diamond knot" when binding some of his victims. This was an important clue, as it is considered a decorative knot used for nautical purposes. This makes sense, as DeAngelo served in the Navy.

Genetic genealogist Margaret Press detailed how this finding of familial DNA can be both confusing and ethically questionable.

It's very controversial . . . It's going to be debated for a very long time in law and forensics and genealogy and everywhere you can imagine. First, how it was not done. Direct-to-consumer DNA testing companies like 23andMe and AncestryDNA did not directly hand customer information over to police. Nor could law enforcement have sent DNA from the crime scene to these two companies, which require a large tube of saliva. Both 23andMe and Ancestry have denied being involved in the case. But customers can themselves choose to export the raw data file from these and other DNA-testing services to a third-party site, such as GEDmatch. These third-party sites are less user friendly than the websites of 23andMe or AncestryDNA, but they offer a more powerful suite of tools. For example, GEDmatch allows users to find profiles that match only one particular segment of DNA. It also lets users who have tested with different services match with each other without shelling out for another one. GEDmatch offers premium tools but is largely free to use.[4]

While some may worry that their DNA is being used to catch their wanted relatives, others would happily enter their DNA in GEDmatch to find those who deserve justice. It is yet another moral dilemma that science and law enforcement have come up against as our collective scientific knowledge grows. Since the apprehension of Joseph DeAngelo, this "long-ranged" familial DNA match has been used to arrest a number of individuals. The first to be convicted, while DeAngelo waited for his own sentence, was fifty-nine-year-old John Miller of Indiana. The unsolved murder of April Tinsely in 1988 had been a horrific crime. This precious child had been raped, strangled, and left in a ditch, and her murderer taunted them with gruesome notes scrawled on barns and other little girls' bicycles as the years wore on. (DeAngelo was also prone to provoking his victims and their families with several taunting phone calls.) Finally, geneticist CeCe Moore found the key to capturing little Tinsely's unknown killer. Journalist Megan Molteni outlined Moore's work for *Wired*:

Eventually she found four couples, born between 1809 and 1849. Once she had them, she could move forward in history, building out family trees of every generation until the present. She did this by tracking names and faces through census records, newspaper archives, school yearbooks, and social media. By the time night fell over her home in San Diego, she had begun to close in on a single branch, into which the four genetic tributaries all ran. From there things moved quickly. As the clock ticked past midnight, she found the relatives that had struck out for Indiana. It didn't take much longer to circle in on two brothers who lived in the area where Tinsley was murdered. Full siblings are as close as genetic genealogy can get. But Moore had a hunch. One brother struck her as a recluse; he had no wife or kids, he lived in a trailer, there were no pictures of him anywhere, and his family never mentioned him on Facebook.[5]

Both Joseph DeAngelo, sentenced to twelve life terms for his atrocities in 2020, and John Miller, sentenced to eighty years in 2018, were apprehended because of long-ranged familial DNA. Law enforcement agencies all over the world are benefiting from this scientific discovery. Especially for cold cases that hold DNA evidence, often left by criminals who could not imagine the advancements of the future.

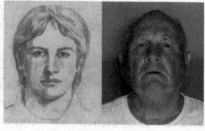

The Golden State Killer alongside a police sketch of him.

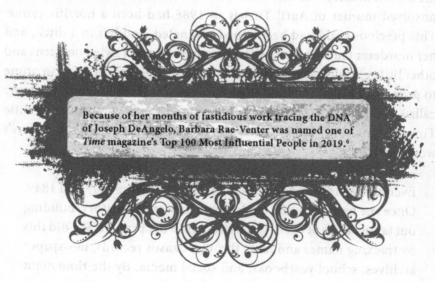

Because of her months of fastidious work tracing the DNA of Joseph DeAngelo, Barbara Rae-Venter was named one of *Time* magazine's Top 100 Most Influential People in 2019.[6]

When Michelle McNamara wrote an article on the Golden State Killer, and then expanded her research into a book, she knew there was hope that he would be caught, that it was a solvable case. It became her obsession to find justice for victims like Jane Carson Sandler, whose life was forever changed by her sexual assault. In 2020, HBO adapted McNamara's book into a documentary series. The final scenes show a victims' gathering. Those who have lost loved ones, those who have been harmed, irreparably, by Joseph DeAngelo's violence, hold hands and rejoice knowing that Michelle McNamara's "Letter to an Old Man" came true. Thanks to her research, the law enforcement's dogmatic persistence, and the savvy of scientists, the doorbell rang.

CHAPTER TWELVE
My Friend Dahmer (Jeffrey Dahmer)

Often, when filmmakers take on a serial killer, especially one as infamous as Jeffrey Dahmer, they focus on the gore. Their murders are reenacted, dissected, and scored with thrilling music to incite fear. While their lives before they descend into evil may be glossed over quickly in a biopic, it is the monster that we see throughout, metaphorical fangs and all.

In *My Friend Dahmer* (2017), we are shown a different side of the man who stalked the streets of Milwaukee. Though Jeffrey Dahmer would later be convicted of seventeen brutal murders, this film, based on the 2012 graphic novel of his childhood friend John Backderf, recounts their friendship in suburban Ohio in the 1970s. It is less about murder and more about the making of a murderer. This new twist on the serial killer film could be far more frightening in its portrayal of the unraveling of a teenager. It is knowing the outcome that makes it all the more tragic.

Serial killer Jeffrey Dahmer.

We've come to know Jeffrey Dahmer as a loner in high school with a rather average home life. There was tension between his parents, but unlike many serial killers, Dahmer was not brought up in a home of intense abuse or abject poverty. Despite the lack of these triggers, there was something peculiar about Jeffrey. He spent much of his time experimenting with found animal bones in a shed, and his usual sexual fantasies were dark and abnormal, as he associated sex with death.

In an attempt to create "zombies," Dahmer bored crude holes into some of his victims' skulls and poured in muriatic acid. Not surprisingly, it didn't work.[1]

On the flip side, in the film we see him find friends in school. He is decidedly "nerdy," yet he appeals to the other students on the fringe, including aspiring artist John Backderf (Alex Wolff). Dahmer (Ross Lynch), Backderf, and their buddies cause mischief at school, getting up to typical teen shenanigans. But, as we know, Jeffrey Dahmer is anything but typical. He killed seventeen men and teenagers and engaged in rape,

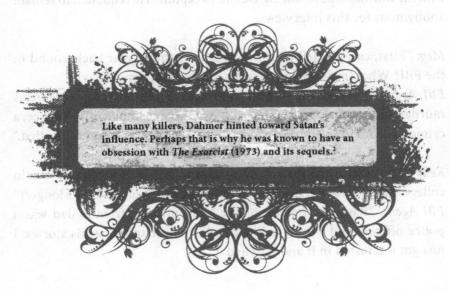

Like many killers, Dahmer hinted toward Satan's influence. Perhaps that is why he was known to have an obsession with *The Exorcist* (1973) and its sequels.[2]

torture, and even cannibalism. The film ends with his first murder of hitchhiker Steve Hicks, just a few weeks after Dahmer graduated from high school. He becomes the Jeffrey Dahmer that we know today.

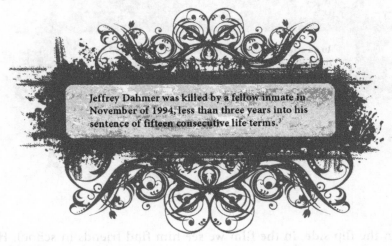

Jeffrey Dahmer was killed by a fellow inmate in November of 1994, less than three years into his sentence of fifteen consecutive life terms.[3]

My Friend Dahmer gives us a unique perspective on a serial murderer. Another viewpoint is from a man who actually hunted and interviewed Jeffrey Dahmer. We had the honor to speak to a retired FBI Agent with an illustrious career spanning over three decades. This includes his time with the Milwaukee FBI during Dahmer's capture. He requested to remain anonymous for this interview.

Meg: "First, can you tell us about your over thirty-year background in the FBI? What inspired you to work in law enforcement?"
FBI Agent: "I worked my entire career in violent crimes, which was murder, kidnapping, bank robberies. Anything dealing with violence; a crime against kids, children, pedophiles, gang murders, things like that."

Kelly: "And what inspired you? Did you know right away when you were in college that you wanted to work in law enforcement or did it take longer?"
FBI Agent: "Basically, I think junior high, my best friend's dad was a police officer. And I used to like to talk to him and hear his stories. I just got interested in it and pursued it."

Meg: "Your work ranged from, like you said, street gangs, bank robberies, to serial killers. Was there a particular type of crime that you found the most fulfilling to solve?"

FBI Agent: "Oh, boy. Well, the serial murders, because they're difficult. Usually the victims are high risk, like prostitutes and runaways. They're hard to solve because there's no connection with the victim and the killer. It's not like domestic homicides, when it's the nearest and dearest, someone they know. It's no secret that husbands kill wives, wives kill husbands. Murders by a stranger can be more difficult so it can be rewarding."

Kelly: "Serial killers are shown as inherently evil in books and films. In your experience, what did you learn about their motives and personalities that might surprise the average person who hasn't been in law enforcement? Someone who has only seen serial killers in the movies?"

FBI Agent: "Well, the vast majority of them are intelligent. They're very well organized and tend to hide in plain sight. People think that we're going to look for monsters, and they don't look like monsters. They can be good looking, and sometimes, very good, like Ted Bundy. Jeffrey Dahmer was very unassuming.

Serial murderers' personalities tend to be similar, and they seem to have common characteristics or events in their background that we've discovered. For one, they have strong fantasies. These fantasies are of violence; torture, binding, suffering, things like that. This leads to excessive masturbation for these people. One thing which reinforces these fantasies is pornography. And it's not like your *Playboy* variety. It's hardcore for them. Also, they generally have a low self-esteem, so they work to increase their self-esteem. They are very lonely people, with little social attachment, and then there's almost always abuse in their background. This can be one of three or a mix; physical, emotional, or sexual. The last thing is a cold or unavailable father figure."

Meg: "Oh, that's interesting. Because, through my reading, it does seem like the mother has a lot of blame placed on her. It's fascinating to hear you say that the father is often detached."

FBI Agent: "Yes, and there are cases where the mother didn't help. We talk about nature versus nurture and there are certain behaviors which people are genetically predisposed to do. But, that doesn't mean everybody that has a certain gene is going to grow up to be Jeff Dahmer. Personality is really the interaction with these things; genetics, psychology, and sociology. You can't pick your parents' genes. So, you may be predisposed to certain behavior. The psychology is how you filter your environment and how you see things. And then the social is your interaction with your peers, your friends. The example I would give is you can have two identical twins, the same genes, same environment, same everything. But one grows up to be an attorney and the other one grows up to need an attorney! And the difference is that third aspect of social interaction. Personality is interaction with those three things."

Kelly: "It seems like the Ted Bundys of the world are very confident, as if they think they're smarter than everyone. It's fascinating to hear that it's kind of a mask like everything else."

FBI Agent: "Yeah. They can have a public persona, that we all have, really. They can put on a very good public persona, but you dig down and they're just lonely people, they constantly need validation. A lot of times you'll see them script their victims. They want their victims to do certain things or say things or act a certain way."

Kelly: "I know that you have vast experience in interviews and interrogation. When you talked to him, did you keep to a particular script? How does it work?"

FBI Agent: "You know, the general approach is always be a friend. I interviewed him four times and I learned a tremendous amount. Like how he got away with it for so long."

Meg: "So, how *did* he get away with it for so long? And in that particular time, maybe the radar wasn't on for these vulnerable victims, like sex workers?"

FBI Agent: "Oh, yes. Dahmer didn't bring attention to himself. He wasn't aggressive. He was just kind of quiet and his victims were often runaways or prostitutes and their disappearance wasn't noticed right away. And by the time somebody did report them missing, it was often months later and very, very difficult then to connect the three things that solve cases: physical evidence, eyewitnesses, and confession. We also need the ability to link the suspect with the victim and the suspect at the crime scene. With serial murders it's difficult because the crime scenes change; they can do the murder, but you don't find out about it for a month. I worked on the Green River Killer Case, Gary Ridgway. Some of those girls, homeless victims, go up to strange men. They could have been dead for two years before the families notice. So, it's like, where do you start? There's no direct link between a victim and the subject. That's why it makes it so difficult to know."

Kelly: "Someone like Gary Ridgeway or Jeffrey Dahmer, they know that these are people are not going to be missed and they're preying on that particular vulnerability."

FBI Agent: "The first question that you have to ask is why this victim on this day at this place? Was this person a bit high risk, like a prostitute or drug dealer? Those types of activities are risky. Or are they a low-risk victim, like a suburban housewife? A child shouldn't be a victim and the rule of thumb is the younger the age of the child, the lower the risk. Unless they grew up in prostituting, you know, or slinging dope. So, when you get a very low-risk victim, you just focus on the circle around them, of people they know. The opposite to a serial killing of a high-risk victim."

Kelly: "Are there any misconceptions about serial killers you have come across?"

FBI Agent: "Yes. Media, television, movies portray it a certain way. But the truth is, serial killers are just pure evil, and most people can't understand that. They can't wrap their head around it; that some people out there can get sexually aroused by hurting another person."

Meg: "How do you grapple with being in a dark place and seeing humanity like you do? How do you not let that color the way you see life in general?"

FBI Agent: "Well, you've got to have a good support system. You've got to have some outside interests. Law enforcement has the highest rate of alcoholism in any profession and also the highest rate of suicide. It's just a tough profession because you see things and you get involved with people who want to hurt you and kill you. That's not fun!"

Kelly: "Have you found over your decades in law enforcement that things have improved? That there is more help in navigating the dark aspects?"
FBI Agent: "Yes! Back in the beginning, you were just kind of expected to 'man up.' If you showed any weakness, they put you on the 'rubber gun squad.' But, as time went on, they realized that there were negative aspects."

It was an absolute honor to speak with someone who devoted his life to hunting down the worst criminals in the country. Instead of focusing on the cruel men and women who take lives, there is inspiration and hope to be found in those who face true horrors to make the world safer. And there is nothing more horrific than what was found in the apartment of Jeffrey Dahmer, or in the confines of his mind.

SECTION FIVE

HIDING IN PLAIN SIGHT

CHAPTER THIRTEEN
Summer of Sam (David Berkowitz)

The summer of 1977 was when I (Kelly) was born. "I'm Your Boogie Man" by KC and the Sunshine Band was number one on the Billboard Hot 100 chart, *Star Wars* was top at the box office, and convicted Martin Luther King Jr. assassin James Earl Ray escaped prison (and was ultimately recaptured). It was also known as the Summer of Sam. David Berkowitz, known as the "Son of Sam," is the notorious serial killer who murdered six people and wounded seven others in New York City. Spike Lee directed the film *Summer of Sam* (1999) and said, in regard to the film's criticisms by the families of the victims, "You know, we understand the pain that they have felt for the last twenty plus years. No one can bring back their loved ones. But number one, this film is not just about David Berkowitz or the Son of Sam. It's about that crazy summer. In no way, shape, or form do we exploit David Berkowitz or the victims in this film."[1]

The blackouts and brownouts of 1977 were prominent features of that hot summer in New York City. How does heat affect us psychologically? First, extreme heat makes people more likely to be cranky. This, in turn, can lead to more temper flares and arguments. One study found that increased heat "leads to a 4 percent increase in interpersonal violence and 14 percent increase in group violence."[2] This could partially explain the events during the twenty-five-hour blackout in July of 1977. Crime rates in New York City were already historically high by this point in the summer and the blackouts led to looting, arson, and at least one murder. Harvard researchers found that cognition and complex tasks like working memory are significantly impaired in extreme heat.[3] But for those with psychological conditions, a heat wave can be extremely dangerous, causing them to lose their ability to care for themselves or have good judgement.

"The United States' power infrastructure is aging rapidly. In 2014, the country's electric grid was losing power more often than any other developing nation. Customers in Japan, for example, averaged four minutes of downtime every year, while customers in the Pacific Northwest averaged about two hundred and fourteen minutes of downtime. The Department of Energy indicated that these power outages could be costing companies as much as $150 billion per year."[4]

There is a famous urban legend of a couple making out in a car when a killer comes to get them. This trope, often seen in horror movies, happened in reality in Berkowitz's case! Berkowitz shot at several pairs of people over the years, some couples and some friends. Because of this pattern, the parents in *The Son of Sam* plead with their children not to go out.

Mira Sorvino's character, Dionna, knows John Leguizamo's character, Vinny, is cheating on her because she smells something on his kiss. Is it possible to recognize other people's scents? Humans have a more powerful sense of smell than we realize. Neuroscientists have found that we can distinguish between one trillion different odors, up from a previous estimate of ten thousand.[5] We can identify people and even

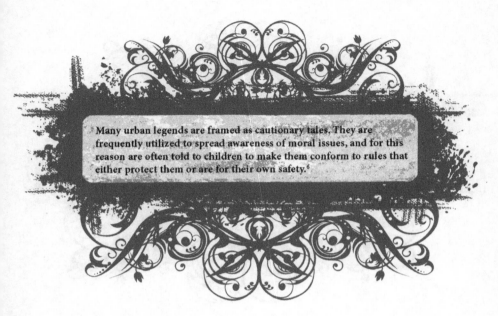

Many urban legends are framed as cautionary tales. They are frequently utilized to spread awareness of moral issues, and for this reason are often told to children to make them conform to rules that either protect them or are for their own safety.[6]

their emotions by scent alone. This means that Dionna could easily have smelled another woman on Vinny and known he was being disloyal.

Speaking of cheating, Vinny seems to cheat a lot. What is the science behind cheating and sex addiction? This topic has been widely debated among experts and some believe that claiming sex addiction as a medical or psychological condition frames people's sex drives and interests into the categories of "normal" or "not normal." The American Association of Sexuality Educators, Counselors, and Therapists released a statement that said "perceptions of sexual 'addictions' may have more to do with people's religious or cultural beliefs than of actual scientific data. The concept of sex addiction emerged in the 1980s as a socially conservative response to cultural anxieties and has gained acceptance through its reliance on medicalization and popular culture visibility."[7] The DSM-5, the reference guide for mental illnesses, does not recognize sex addiction because it does not change the brain like other addictions. Others believe that the science in this field is still emerging and that medical professionals may recognize it in the future.

David Berkowitz wrote letters to the police that helped identify him. The science of handwriting identification and handwriting analysis has been around since the time of Aristotle. Graphology, identifying traits

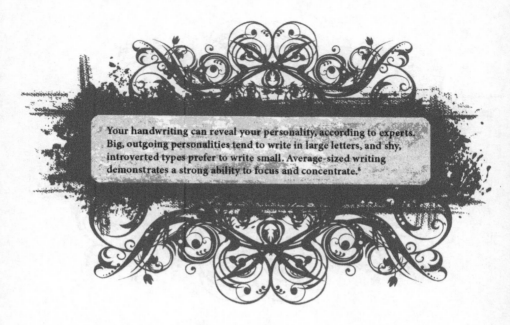

Your handwriting can reveal your personality, according to experts. Big, outgoing personalities tend to write in large letters, and shy, introverted types prefer to write small. Average-sized writing demonstrates a strong ability to focus and concentrate.[5]

through handwriting, is not only used in criminal investigations but also in analyzing health, personality, and compatibility in relationships. Like fingerprints, no two writers will share the exact same combination of handwriting characteristics. The *International Journal of Criminal Investigation* explains, "From a graphologist's point of view, the writing of a psychopath can generally be described as relatively conformist, rather banal, with little rhythm, stiff, monotonous, and abounding in abnormalities." The *Journal* goes on to explain Berkowitz's writing:

The signs of abnormality are deceptively more subtle, though equally important in the analysis. However, they point toward a disturbed,

David Berkowitz left a letter for the police near two of his victims. "The letter expressed the killer's determination to continue his work, and taunted police for their fruitless efforts to capture him."[9]

easy-to-influence person, with a rather narrow concept of life and a genuine incapacity to create and sustain healthy relationships, who suffers from a powerful complex of inferiority and who is incapable of seeing the connection between his deeds and their outcoming results.[10]

When reports surfaced that Berkowitz was being offered money for the rights to his story, the New York Assembly passed the "Son of Sam" law, requiring that "an accused or convicted criminal's income from works describing his crime be deposited in an escrow account. The funds from the escrow account were then to be used to reimburse crime victims for the harm they had suffered."[11]

David Berkowitz liked women with long, dark hair. When this information was made public, women went as far as changing their hair color or wearing wigs so that they wouldn't look like "his type." What is the science of attraction? Our opinions about the attractiveness of various hair colors, natural or otherwise, are drawn from a combination of historical reference, cultural prejudices, possible ideas about rarity and disease protection, and a bunch of other factors."[12]

Berkowitz was known as the ".44 Caliber Killer" because of the gun he used. TheTrace.org explains the science behind bullets and the damage they cause:

Projectile weapons work by transferring kinetic energy to a target, which ripples out as a shockwave through tissue as the bullet plows through the body, leaving a cavity in its wake. The amount of energy a bullet radiates into a target is determined by a simple formula taught in high school: It's the product of one half the projectile's mass times the square of the velocity. The energy delivered to the target increases geometrically along with increases in mass, and exponentially with increases in velocity. The larger a projectile's surface area, the greater its ability to transfer its energy to the target, instead of simply penetrating straight through."[13]

Although not all shooting victims die, both physical and emotional wounds take time to heal. A study in 2019 revealed that many gunshot victims suffer increased unemployment, drug and alcohol abuse, as well as post-traumatic stress disorder.[14]

Berkowitz enjoyed the feeling of holding "social power" over police, the media, and the general public. There are several fictional horror villains that fit into this definition including Jigsaw from *Saw* (2004) and John Doe from *Seven* (1995). What is the psychology behind this? Dr. Scott Bonn, who met with Berkowitz in 2013, said "it is clear to me that Berkowitz relished his evil celebrity status and that he enjoyed terrorizing the city of New York throughout his murderous rampage. I believe that his criminal infamy boosted his otherwise fragile, disturbed ego and gave him a twisted sense of identity and purpose."[15] This extreme example of social power proves how far this feeling can progress, but for most people, a sense of power is a normal, everyday interpersonal interaction. Being aware of the power we hold, and the power others hold, helps us to understand our relationship dynamics.

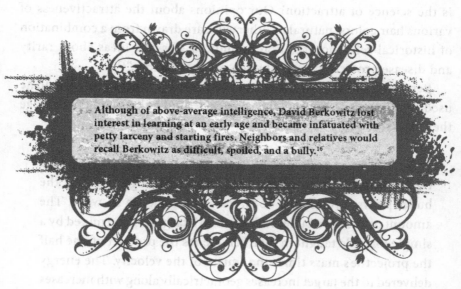

Although of above-average intelligence, David Berkowitz lost interest in learning at an early age and became infatuated with petty larceny and starting fires. Neighbors and relatives would recall Berkowitz as difficult, spoiled, and a bully.[16]

David Berkowitz in *Summer of Sam* has a break with reality and begins to see a black dog that talks to him. His lawyers wanted to use this story in an attempt to mount an insanity plea, but he refused. What is the science behind hallucinations? Surprisingly, hallucinations are quite common. One in twenty people report hearing or seeing things that others don't while they're fully awake.[17] Hallucinations are caused by miscommunication between the brain's frontal lobe and the sensory cortex. Not all hallucinations are negative or scary; most tend to be completely benign.

People got sick of being patient and staying inside during that summer of 1977. They started going out with abandon, similar to coronavirus quarantine fatigue. The long-term effects of the COVID-19 pandemic won't be known for some time, but the parallels are striking.

CHAPTER FOURTEEN
Dear Mr. Gacy
(John Wayne Gacy)

When many of us hear the name John Wayne Gacy, we picture the sinister figure of a pale clown face with an exaggerated red mouth and bright blue eyes. This twisted serial killer dressed up as a clown for children's birthday parties and some experts believe that this act allowed him to regress into childhood. "Sigmund Freud believed age regression was an unconscious defense mechanism. It was a way the ego could protect itself from trauma, stress, or anger. Still, other psychologists think of

POLICE DEPT.

John Wayne Gacy "regularly performed at children's hospitals and charitable events as 'Pogo the Clown' or 'Patches the Clown,' personas he had devised."[2]

age regression as a way for people to achieve a therapeutic goal. It might be used to help a patient recall memories of trauma or painful events."[1] Regression may be a symptom of a greater psychiatric issue including schizophrenia, dissociative identity disorder, dementia, or borderline personality disorder. Gacy murdered at least thirty-three young men and boys and has become an infamous figure in American history.

The film *Dear Mr. Gacy* (2010) is based on the book *The Last Victim* (1999) by Jason Moss, who interviewed Gacy extensively, along with other notorious serial killers. We had the opportunity to talk to filmmaker Kellie Madison, who wrote the screenplay for *Dear Mr. Gacy*, and learned some fascinating things:

Kelly: "How much did you personally know about John Wayne Gacy before you took on this project?"

Kellie Madison: "Well, girl, I'm going to blow you away with this one! I actually grew up on John Wayne Gacy's block. I literally lived at 8611 Summerdale and he killed at 8213 West Summerdale so my mom used to tell me and my brother, walking past their property every day, 'there's bad people in the world and you should never talk to strangers' and we'd get the whole speech because 'a bad man did horrible things here.' So, we knew about John Wayne Gacy very, very early on. That's how I got enculturated to Mr. Gacy, so I think there's a strong connection there from my childhood. In my twenties I got into true crime, and I started reading every true crime book imaginable, and I came across *The Last Victim*. I was riveted by this book."

Meg: "Going back to growing up in that world and your mom warning you, do you think that shaped you as a person? Do you think you were more careful because you thought there could be a serial killer among us?"

Kellie Madison: "I think as long as I can remember I've been absolutely terrified that everyone I meet is a serial killer. Because I've got so much proclivity toward the dark side it makes me be extra cautious. There's a couple of time periods in my life when I actually had to take a break from reading and watching dark things because it does get into your subconscious way too much. But I do happen to be fascinated by the perversion of the human mind and how much of it is nature versus nurture and how much of it is brain damage. I'm so fascinated by the criminal mind.

Kelly: "You mentioned that you read the book first. How did you get on this project? Did you seek it out?"

Kellie Madison: "I found the book and then I hunted down the author, who was Jason Moss, and I contacted him myself. He was a practicing criminal defense attorney in Las Vegas at the time. He said the option rights had just become available, so I then contacted another producer that was bigger than I was, because this was virtually my first film, and that producer was Clark Peterson. He had just come off of the movie

Monster (2003) with Charlize Theron. I asked if he was open to hearing a pitch about another true crime or serial killer movie. He was. It took six months to negotiate the contract. We were just about to close on the deal and Clark called me up and said 'Kellie, are you sitting down?' and he said 'Jason Moss just shot himself in the head.'"

Kelly: "Wow."

Kellie Madison: "He shot himself on 666 [June 6, 2006] so he had obviously planned it and I was devastated. I was the only one who had interacted with Jason on our team. There must have been such grave, drastic, dramatic things going on in his life to lead up to that point."

Meg: "How awful."

Kellie Madison: "It was only at that point in time that we found out he had a wife. Charlotte Moss was the one who retained the book rights and the life rights. We flew her to LA because we wanted to sit down with her and interview her about Jason. What happened from the time he visited John Wayne Gacy in prison in 1994 to when he decided to shoot himself?"

Kelly: "What did she say?"

Kellie Madison: "I was on the edge of my seat listening to her talk about this process of what Jason went through. Jason got so fascinated, in a negative way, with serial killers. She [Charlotte] said he was following the serial killer trajectory."

Meg: "When you were writing this script and going through the book, how did it affect you personally?"

Kellie Madison: "It's always going to impact you; subconsciously, subliminally. You start to get more scared everywhere you go. I went to some dark places with Gacy for sure. I went down to the prison he was in and I spoke to a lot of the cops who were on the case. I really got into the world by interviewing them for hours. I'd be writing in my house, and it's dark, and then a noise [gasps], and I'd literally jump because I'm in this space of darkness. It can't help but affect you."

Kelly: "I've felt that way even just doing research for this book! Sometimes I need to step away."

Kellie Madison's interview certainly gave us a new perspective when rewatching this film and revisiting this case.

The childhood relationship that Gacy had with his father may have contributed to his psyche. According to experts, "a large body of research on attitudes indicates that parental warmth together with reasonable levels of control combine to produce positive child outcomes."[3] Gacy's father drank heavily and often demeaned him. He was berated for being clumsy and was beaten along with his mother. As with most instances of abuse, a person's history may explain their outlook or behavior, but it doesn't excuse it.

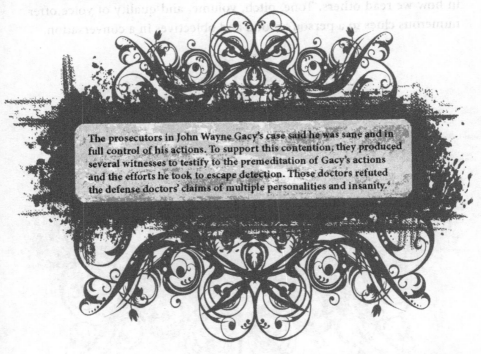

The prosecutors in John Wayne Gacy's case said he was sane and in full control of his actions. To support this contention, they produced several witnesses to testify to the premeditation of Gacy's actions and the efforts he took to escape detection. Those doctors refuted the defense doctors' claims of multiple personalities and insanity.[4]

Gacy was said to have antisocial personality disorder. According to the Mayo Clinic, "antisocial personality disorder, sometimes called sociopathy, is a mental disorder in which a person consistently shows no regard for right and wrong and ignores the rights and feelings of others. People with antisocial personality disorder tend to antagonize, manipulate, or treat others harshly or with callous indifference. They show no guilt or remorse for their behavior."[5] This absolutely describes multiple serial killers we cover in this book and explains their behavior. Many films feature characters that have this condition including *The Godfather* (1972,) *Silence of the Lambs* (1991,) and *There Will Be Blood* (2007).[6]

John Wayne Gacy tells Jason in the film *Dear Mr. Gacy* to observe people to understand them better. Facial expression, posture, eye contact or lack thereof, and spatial distance all reveal something about a person. We can derive details about someone's personality without ever hearing them speak. Speech, though, is also a part of nonverbal communication. The vocal aspects of language, not the words themselves, play a part in how we read others. Tone, pitch, volume, and quality of voice offer numerous clues to a person's mood and objectives in a conversation.

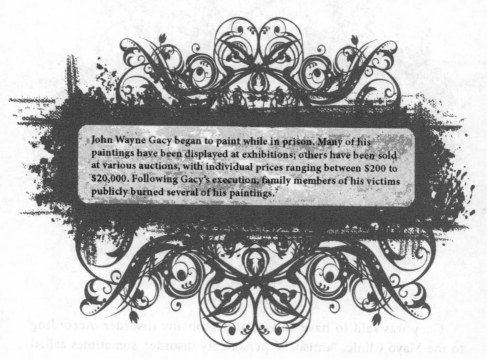

John Wayne Gacy began to paint while in prison. Many of his paintings have been displayed at exhibitions; others have been sold at various auctions, with individual prices ranging between $200 to $20,000. Following Gacy's execution, family members of his victims publicly burned several of his paintings.[7]

Gacy repeatedly drowned then revived his victims. This monstrous act is rooted in our human psyches as a primal terror. We are not meant to be submerged in water, unable to breathe. When we are underwater, we

naturally hold our breath. But when time passes, we inadvertently gasp and take in water. If the inhalation of water continues, organs will fail. "In 1776, doctor and inventor John Hunter proposed a double bellows to breathe air in and out of the lungs like a fireplace bellows, and remarkably similar to the positive-pressure ventilation used in modern respirators."[8]

Gacy used quicklime to hasten decomposition in his victims' bodies. How does it work? "Quicklime is calcium oxide. When it contacts water, as it often does in burial sites, it reacts with the water to make calcium hydroxide, also known as slaked lime. This corrosive material may damage the corpse, but the heat produced from this activity will kill many of the putrefying bacteria and dehydrate the body."[9] Quicklime isn't always used for clandestine reasons. In the Iron Age, quicklime burials were the norm in Spain. Currently, it has a wide range of uses, including the production of iron and steel, paper and pulp production, treatment of water and flue gases, and in the mining industry.[10]

Alkaline hydrolysis (also called aquamation, biocremation, resomation, flameless cremation, or water cremation) is a process for the disposal of human and pet remains using lye and heat. The process is being marketed as an alternative to the traditional options of burial or cremation.[11]

When children go missing today, an Amber Alert is issued. The Amber Alert system is an acronym for America's Missing: Broadcast Emergency Response. The alert was named after Amber Hagerman, a nine-year-old girl abducted and murdered in Arlington, Texas, in 1996. How is this related to John Wayne Gacy? His defense attorney, Sam Amirante, helped establish it. *The Lineup* explains:

> In 1984, Amirante created the Missing Child Recovery Act. Prior to this legislation, Illinois authorities had to wait seventy-two hours after a child was reported missing before beginning a search. Amirante's legislature removed this . . . waiting period, allowing authorities to begin searching right away. Other states adopted similar laws, and a nationwide network for missing children was gradually formed, today known as the Amber Alert System.[12]

The psychology of posing as a potential victim had a tremendous strain on Jason Moss and his life. The movie is hard to watch, knowing his actual fate. Was the date of 6-6-6 for Jason's suicide a coincidence or a plan? We may never know for certain, but it seems Moss truly was Gacy's last victim.

CHAPTER FIFTEEN

A Good Marriage
(The BTK Killer)

Stephen King wrote "A Good Marriage" for his book *Full Dark, No Stars* (2010) after reading an article about the BTK killer, Dennis Rader, who murdered ten people between 1974 and 2001. Rader left clues behind to mock the police, which eventually led to his downfall. Metadata from a crime scene was used to track him down and he is currently serving ten life sentences without parole. King said:

Dennis Rader was known as the BTK Killer, an abbreviation he gave himself, which stood for "bind, torture, kill."

Paula Rader was married to this monster for thirty-four years, and many in the Wichita area, where Rader claimed his victims, refuse to believe that she could live with him and not know what he was doing. I did believe—I do believe—and I wrote this story to explore what might happen in such a case if the wife suddenly found out about her husband's awful hobby. I also wrote it to explore the idea that it's impossible to fully know anyone, even those we love the most.[1]

Some secrets we keep are trivial while others may be criminal. Robert Motta, director of the doctoral program at Hofstra University's School of Community Psychology and a psychologist in Hempstead, New York, said, "leading a double life is not as uncommon or abnormal as it may sound. According to recent statistics, 70 percent of all males and 50 percent of all females are going to have an extramarital affair at some point over the course of their marriage. What this says is that most people, at some time in their life, are going to lead double lives."[2] What is it that makes certain people feel compelled to keep secrets or live double lives? Some psychologists suggest that it lies in people's personalities. If someone has conflicting parts of their personality, they are more likely to compartmentalize and use lying as a strategy to cope. It's not uncommon for healthy, well-adjusted people to have a public self and a private self, though. We may act differently in a work setting than we do at home. We may highlight or diminish certain aspects of our personalities based on the people we're around. It only becomes dangerous if the actions are high-risk, extreme, or irrational.

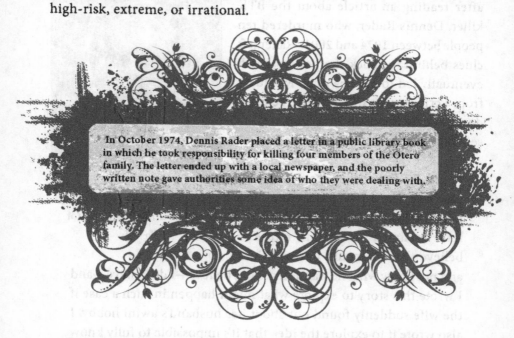

In October 1974, Dennis Rader placed a letter in a public library book in which he took responsibility for killing four members of the Otero family. The letter ended up with a local newspaper, and the poorly written note gave authorities some idea of who they were dealing with.[3]

What are the laws regarding spouses knowing about their partner's crimes? Spousal privilege protects the confidential communication between spouses and protects spouses from having to testify against one another. Although the wives in the movie and in the real-life case of Dennis Rader knew nothing of their spouse's crimes, they would not have been legally obligated to testify in court. Rader's daughter, Kerri Rawson, told the *Wichita Eagle* that there was "no way" her mother could have known about her father's serial killer side, adding, "she wouldn't have raised us with him."[4]

The movie *A Good Marriage* focuses on Darcy (Joan Allen), the wife of Bob (Anthony LaPaglia), a serial killer. Darcy finds Bob's "trophies" from his victims in a hidden container. This is like the real BTK killer keeping the driver's licenses of his victims. Do many killers keep tokens or mementos from their crime scenes? According to criminal profilers, taking items "helps prolong, even nourish, their fantasy of the crime."[5] Serial killer Ivan Milat (page 39), kept the camping supplies of his victims while Anatoly Onoprienko kept the underwear of his fifty-two victims. In the film, after she discovers her husband's horrifying secret, Darcy grapples with the fear of coming clean and her life changing. Similar to being worried about leaving an abusive relationship, she was choosing to live with the devil she knows instead of the devil she doesn't. She promises to keep his secret but asks him to promise to stop killing. He claims he will.

German serial killer Fritz Haarmann, known as the "Vampire of Hanover" murdered at least twenty-four young men between 1918 and 1924. He would collect various items of his victims' belongings to take home with him. He was known to sell these items for money or give them away to friends and family as gifts, all the while having the knowledge of their true origins. Haarmann was sentenced to death for his crimes and executed in 1925.[6]

Darcy begins to imagine newspeople talking to her on TV. Are hallucinations common during times of extreme stress? According to health professionals, "while it's rare for someone with anxiety to truly hallucinate, it's not rare for those with intense anxiety to have various types of mild hallucinations that can cause additional fear over your mental stability."[7] It's easy to understand, then, why Darcy would experience these strange occurrences. Some other types of hallucinations that those with extreme anxiety may experience include visual or light changes, auditory or daydream sounds, and even olfactory hallucinations such as smelling something that isn't there. Experts recommend treating your anxiety with a health-care professional and attending therapy if advised.

Rader was said to have an obsession with zoosadism, which is the act of torturing, killing, and hanging small animals. What is the psychology behind this condition? The FBI found that "a history of cruelty to animals is one of the traits that regularly appear in its computer records of serial rapists and murderers, and the standard diagnostic and treatment manual for psychiatric and emotional disorders lists cruelty to animals as a diagnostic criterion for conduct disorders."[8] This act in childhood is considered one of the three that make up the "Homicidal Triad." The other two acts are persistent bed-wetting and obsessive fire-setting.

Rader sent letters to the police as well as a floppy disk that contained metadata for the phrase "Christ Lutheran Church" and the name "Dennis." This ultimately led to his arrest and conviction. Metadata collection has advanced far since then. Not only can investigators tell who we talk to and for how long on our cell phones but can also see exactly where we are when a call takes place. In a paper published in *Nature's Scientific Reports*, "MIT researchers found that with cell phone call metadata from 1.5 million anonymous people, they could identify a person easily with just four phone calls. As Foreign Policy's Joshua Keating explains, they didn't need names, addresses, or phone numbers. They only used the time of the call and the closest cell tower."[9] Metadata can reveal even more information about us including our hobbies, interests, and social interactions. This advance in technology may feel worrisome to the common person but must be terrifying for potential criminals.

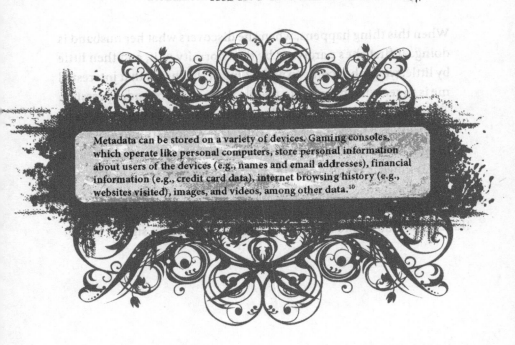

Metadata can be stored on a variety of devices. Gaming consoles, which operate like personal computers, store personal information about users of the devices (e.g., names and email addresses), financial information (e.g., credit card data), internet browsing history (e.g., websites visited), images, and videos, among other data.[10]

A pap smear from Rader's daughter was used to connect DNA found under a victim's fingernails to Rader himself. How does this work? During any type of assault, DNA evidence may be passed from the perpetrator to the victim. If gathered in an investigation, it can be used to identify potential criminals. The ethics of using a family member's DNA to implicate a relative is hotly debated. In some cases, people submitting their samples to genealogy studies have inadvertently led authorities to a guilty family member. Because they didn't give consent for their information to be used for other purposes, like police investigations, some experts recommend that there is an opt out option for genetic tests. Some law enforcement and lawyers think that if more people submit to testing, more criminals will be caught.

Unlike the real fate of the BTK killer, Darcy kills her husband in *A Good Marriage* by pushing him over the upstairs bannister. She suffocates him with a towel wrapped in a baggie, then rinses the baggie and puts it in the freezer with some hamburger meat. She plays the grieving widow at the funeral even though we, the audience, know the truth. Stephen King, in a 2014 interview, said:

When this thing happens, when she discovers what her husband is doing, at first she's paralyzed by the enormity of it, and then little by little, we see her come to grips. One of the things that interested me is to see what happens to people when they are under pressure. That's what this movie is really about. Joan carried it, and, to be fair, Anthony LaPaglia's a great Bob. There's something going with his portrayal of Bob. You say to yourself, 'This is probably what a serial killer is really like.' He's sort of this ordinary guy with this monster inside.'[11]

Rader, like Bob in the film, truly was hiding in plain sight.

SECTION SIX
I WANT TO BE FAMOUS

CHAPTER SIXTEEN
Scream
(The Gainesville Ripper)

The line between fiction and reality has long been blurred. Our collective understanding of serial killers is often colored by films, TV, music, even comic books. While we know humans like the Gainesville Ripper exist, we are more likely to have spent time with hulking, mute slashers like Jason Voorhees.

In March of 1994, screenwriter Kevin Williamson watched an episode of the TV newsmagazine series *Turning Point* (1994–1999) about Danny Rolling, known as the Gainesville Ripper. The show took an in-depth look at the series of murders perpetrated by Rolling in 1990 in Gainesville, Florida. These gruesome attacks on college coeds inspired Williamson to write *Scream* (1996). *Scream* would go on to be one of the most iconic horror franchises ever, spawning four sequels (*Scream 5* is slated for a 2022 release) and an MTV series (2015–2017). One unique aspect of *Scream* that resonated with audiences was the teenagers' rather impressive knowledge of horror films, from *Halloween* (1978) to *A Nightmare on Elm Street* (1984). This led characters like Randy Meeks (Jamie Kennedy) to point to horror clichés within the *Scream* universe, even changing their behavior to avoid certain death. This dichotomy of reality and fantasy can further muddy how we perceive real killers like Danny Rolling.

The "ghostface" mask gained popularity after the movie *Scream* was released.

The truth is, Danny Rolling was a seriously disturbed killer who wreaked havoc on the University of Florida campus in the fall of 1990. His shadow loomed so large that many students left campus and attended different colleges to remove themselves from the chaos. Some who remained avoided being alone during Rolling's murder spree, and even slept in large groups. Rolling did not wear a ghostface mask, or terrorize his victims on the phone, but the fear and brutality in *Scream* was on point.

What had terrified them so immensely was not only the seemingly random murders, but also the vicious and gruesome way in which they were carried out. Danny Rolling, known that fall as the anonymous Gainesville Ripper, would leave his victims in bizarre poses, spending time to stage them in twisted and often provocative ways. In one case, he placed the victim's detached head on a shelf so that it was turned to "look" upon its own body.

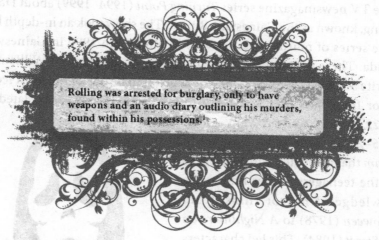

Rolling was arrested for burglary, only to have weapons and an audio diary outlining his murders, found within his possessions.[1]

At the end of *Scream*, the two killers reveal that their motive is to be famous. They have seen how Sydney Prescott (Neve Campbell) was given attention over the murder of her mother, and they have long consumed the media's fixation on murder and mayhem. This theme is abundant in the *Scream* films, as several of the sequels revolve around the movie-within-a-movie theme, and in *Scream 4* (2011) Sydney's cousin Jill (Emma Roberts) reveals herself to be the murderer for the same reasons.

While Danny Rolling murdered long before social media, there is something performative about the brutal tableaus he left behind. Psychologist Dr. Scott A. Bonn describes the difference between "staging" and "posing," which is what Danny Rolling did with his victims:

> The FBI profiler may also encounter deliberate alterations of the crime scene or the victim's body position at the scene of the murder. If these alterations are made for the purpose of confusing or otherwise misleading criminal investigators, then they are called staging and they are considered to be part of the killer's MO. On the other hand, if the crime scene alterations only serve the fantasy needs of the offender, then they are considered part of the signature and they are referred to as posing. Sometimes, a victim's body is posed to send a message to the police or public. For example, Jack the Ripper sometimes posed his victims' nude bodies with their legs spread apart to shock onlookers and the police in Victorian England.[2]

The phenomenon of serial killers posing their victims further illustrates their need for control, as well as an ability to both prolong their fantasy and send a message. A more recent example is Deangelo Kenneth Martin of Detroit, Michigan. (It is important to note that at the time of this writing, Martin was charged but not convicted of these crimes). Martin is accused of murdering three sex workers and posing each on their knees in a sort of prayer style, all in vacant homes. Forensic psychologist Dr. Stephen Raffle evoked Jack the Ripper again when asked what the kneeling could signify:

> What's the symbolism of the kneeling? I presume he's punishing them for their behavior. It's not so different from Jack the Ripper, who also singled out prostitutes as victims. By posing them kneeling, the killer is likely demonstrating to himself and whoever he thinks his audience is that these are fallen women. He's putting himself in God's role as the enforcer of morality. Serial killers, believe it or not, see themselves as highly moralistic people, who are offended at the act of prostitution, to such a degree that

they act out this rage. They want women to be Madonnas, and when a woman is not a Madonna, then that woman is a personal insult to them. So, they punish the transgressor. They have to see women as perfection and nurturing, not as sexual objects. Their own upbringing usually leads to a splitting of good mother/bad mother, and the prostitute represents the bad mother, who must be punished.[3]

The Madonna-whore complex is a concept first described by Sigmund Freud in which a person can only place women in two distinct boxes. They are either a perfect, maternal figure or a debased, perverted whore. This is a trope seen in all aspects of entertainment, often depicted in the films of Alfred Hitchcock. In horror films like *Scream*, the "final girl" is typically in stark contrast to her murdered best friend (Rose McGowan) who is more open with her sexuality. Those who operate with this ill-conceived notion, of course, miss the nuance of personality, unable to understand that a woman can be both or neither of these extreme roles. For more about female tropes as they are depicted in film, TV, and literature, refer to our book *The Science of Women in Horror* (2020).

After his conviction for the murder of five college students, Danny Rolling made a strange life for himself on death row. He met and began a romantic relationship with true-crime writer Sondra London. They were even engaged to be married despite his impending death. While in prison, he wrote poetry and sketched, and because of Rolling's sensationalized case, these artistic items became a hot commodity.

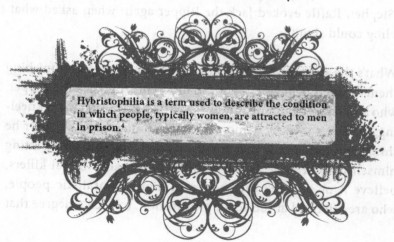

Hybristophilia is a term used to describe the condition in which people, typically women, are attracted to men in prison.[4]

Murderabilia is a term used to describe collectibles related to homicide and violent crimes. It has caused a raucous debate, as some see it as a killer's right to free speech, yet others find it appalling that a murderer could profit off their crimes (because their notoriety is what drives up the price). The Son of Sam Law, created in the 1970s to preemptively prevent David Berkowitz from selling his story, was the first law of its kind. Originating in New York, the law has found its way to other states, and has been revised many times. Detractors of the law argue that monetary incentives could actually be helpful in getting full confessions, and therefore helpful in finding bodies and bringing closure to victims' families.

On the other hand, many victims' families find the profiting of murder distressing. "The whole thing adds insult to injury," said Mario Taboada in an interview with the *Gainesville Sun*.[5] Mario's brother Manny was killed by Danny Rolling when Rolling found him sleeping outside the room of Tracy Paules to protect her from the serial killer on the loose. Mario Taboada was specifically addressing the six-thousand dollars deposited to Rolling's prison account over a five-year period. These were payments for his poems and art, which he could use in the prison commissary to buy snacks and comfort items.

In some cases, the fascination with murderabilia can bring money to those who lost loved ones. In 2011, the US Government auctioned off items from the remote cabin of the Unabomber, Ted Kaczynski. Collectors spent over two-hundred thousand dollars to grab a piece of murder history, and all the money went directly to victims of Kaczynski.

While not everyone has the money or interest in owning such a macabre trinket, many can get a peek of murderabilia at the Museum of Death in Los Angeles and New Orleans. Here, there are rooms filled with serial killers' possessions. For instance, there is a Charles Manson exhibit which includes clothes he wore and a guitar he strummed on. There are also clown shoes worn by John Wayne Gacy, and black and white photographs of the Black Dahlia murder victim, Elizabeth Short.

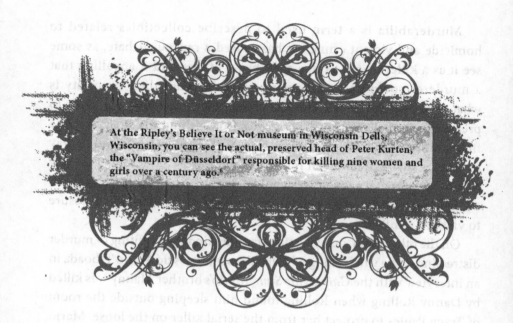

At the Ripley's Believe It or Not museum in Wisconsin Dells, Wisconsin, you can see the actual, preserved head of Peter Kurten, the "Vampire of Düsseldorf" responsible for killing nine women and girls over a century ago.[6]

Why are we inherently fascinated by these items? In his book *Why We Love Serial Killers* (2014), Dr. Scott Bonn explains, "some believe 'items once held by the likes of Bundy or Gacy are endowed with magical powers,' and the owners of such objects—be they museums or private collectors—can tap into that supernatural force. It's called the talisman effect."[7]

A talisman is defined as an object that holds religious or occult-like power that brings protection or healing to the person holding it. The talisman effect is a person's belief in the object's supposed powers. This term is often used in the medical field, as there is thought to be a placebo effect when a patient believes that a medicine holds the key to their health. They often become healthier not only because of the medicine, but because they have hope and positivity. While that seems a far cry from the creepy objects in the Museum of Death, the concept remains similar. By being around these items once touched by humanity's worst monsters, perhaps we can process the undefinable dread of death.

Danny Rolling was executed by lethal injection in October of 2006. His name does not hold the same notoriety as Ted Bundy or Jeffrey Dahmer, yet his murders were equally as depraved and vicious. Like

Ghostface in the *Scream* franchise, he is better known as his moniker, the Gainesville Ripper. Is this our way of stripping away his humanity? Do we want to distance ourselves from who he was? Or, like those who collect clown paintings by Gacy or poems by Rolling, are we drawn to the darker side? In an interview with the *Los Angeles Times* about the Museum of Death, criminal prosecutor Stephen Kay puts it simply: "people like to be scared."[8]

CHAPTER SEVENTEEN
Rope (Leopold and Loeb)

Many serial killers operated for decades before being caught by authorities. Men like Dennis Rader (BTK) and Samuel Little, considered to be some of the most prolific serial killers to date, took advantage of the shadows. They hid in plain sight, satisfying their depraved sexual fantasies at night while wearing a facade of normalcy in the light of day.

This chapter delves into the "wannabes." These are the killers who took their victims' lives because they wanted to feel intellectually superior. They wanted to bask in the glow of their kill, while frustrating the detectives on the case. The irony, of course, is that these men failed at living up to their own inflated egos. While they wanted to emulate BTK, or a murderer never caught like the Zodiac, they instead overestimated their murderous skills.

In Alfred Hitchcock's film *Rope* (1948) Brandon Shaw (John Dall) and Phillip Morgan (Farley Granger) believe they know exactly how to get away with murder. The film, experimental in its use of long, continuous shots, begins with Shaw and Morgan's murder of David Kentley (Dick Hogan). After strangling David with a rope, the duo hides his body in a large chest, just in time for a party. They invite David's family, friends, and even his fiancée, to privately gloat over their achievement. The only motive for this murder is their warped sense of the world. Shaw and Morgan see themselves as superior to others, and by accomplishing this murder just beneath the noses of others, they feel this is proof of their god-like status. Too bad for them, their wise professor Rupert Cadell (James Stewart) notices the cracks between Shaw and Morgan. Morgan is the weak link. He begins to show his nerves in front of the party guests, something Cadell picks up on instantly. And there is the matter of the chest. In his bravado, Shaw insists that the buffet be served on top of the chest in which David's body is hidden. This strange detail, along with Cadell finding David's hat in the hall closet, leads him to realize that a murder has occurred. In classic cinematic style, Cadell nearly dies in his pursuit of the truth, and in the end, Shaw and Morgan are held accountable for their pointless crime.

Rope was based on a play of the same name by playwright Patrick Hamilton, who also wrote the play *Gaslight* (1938), which was turned into a film by Hitchcock in 1944. *Gaslight* is where the term "gaslighting" originated. Like many playwrights, Hamilton was inspired by true events in his creation of *Rope*. In fact, there have been four films based on the same case: *Rope, Compulsion* (1959), *Swoon* (1992), and *Murder by Numbers* (2002). All involve the same basic premise of two young, intellectual men working together to commit the "perfect murder."

So, who were the real men who inspired these films?

Nathan Leopold and Richard Loeb were born into the affluent society of Chicago at the beginning of the twentieth century. Although they didn't spend much time together in childhood, they had a lot in common. Both the sons of wealthy parents, they were also blessed with intelligence. At nineteen, Leopold had already achieved his undergraduate degree from the University of Chicago. He spoke fifteen languages and was a renowned ornithologist. Eighteen-year-old Loeb had skipped several grades in his youth and was known at the time as the youngest graduate of the University of Michigan.

Leopold and Loeb were not the only killers to point to philosophers as inspiration for their crimes. Melvin Rees, killer of at least five during the 1950s, pointed to Jean Paul Sartre's ideas of free will as an excuse to murder.[1]

While both Leopold and Loeb were surely poised for greatness, they used their genius for evil. Fast friends at the University of Chicago, where Loeb was working on post-graduate studies in history, the fusion of Leopold and Loeb was no less harmful than an exploding bomb. One spark to the fuse was both men's interest in the work of Friedrich Nietzsche. German philosopher Nietzsche was known for his rather provocative ideas. Brittanica.com details one such idea known as "perspectivism:"

> Perspectivism is a concept which holds that knowledge is always perspectival, that there are no immaculate perceptions, and that knowledge from no point of view is as incoherent a notion as seeing from no particular vantage point. Perspectivism also denies the possibility of an all-inclusive perspective, which could contain all others and, hence, make reality available as it is in itself. The concept of such an all-inclusive perspective is as incoherent as the concept of seeing an object from every possible vantage point simultaneously.[2]

Basically, perspectivism is the idea that there is no absolute truth because every human perceives an event through varying perspectives. In the minds of young men like Leopold and Loeb, perspectivism became a weapon. They believed that there really is no true right or wrong, despite the morality they were taught in school or church. Another concept explored by Nietzsche was that of Übermensch, or "Superman." In his book Thus Spoke Zarathustra (1883) Nietzsche utilized the fictional character Zarathustra to symbolize the Übermensch. It is a rather complicated concept that he wrote about in various works, an archetype not unlike a hero, who is above or better than the average human. Psychiatrist Dr. Eva Cybulska explains what an Übermensch is:

The free-spirited Übermensch would not succumb to the herd mentality and become a nonentity in some monstrous super-state. Released from the chains of tradition and ideology, such an individual would be free to create new values with a sense of uniqueness and passion for life. In the mood of Greek agon, he would go beyond (über) the narrowness of national divisions and parochial resentments.[3]

Again, teens Leopold and Loeb, and their fictional counterparts, misinterpret Nietzsche's intent. Cybulska explains what an Ubermensch is *not* in order to demonstrate the fallacies of those reading his work. "Übermensch is not a tyrant. If anything, he is someone capable of tyranny who manages to overcome and sublimate this urge. His magnanimity stems not from weakness and servitude, but from the strength of his passions. It's important to stress that there has never yet been an Übermensch; it remains an ideal."[4]

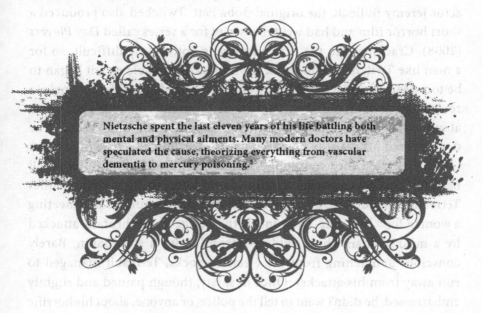

Nietzsche spent the last eleven years of his life battling both mental and physical ailments. Many modern doctors have speculated the cause, theorizing everything from vascular dementia to mercury poisoning.[5]

In *Rope*, Cadell discusses the "Superman" philosophy with his former students, Shaw and Morgan, unaware that they had already seized upon this concept in order to justify their homicide. To them, they are proving themselves to be Supermen in Nietzsche's construction by embracing anarchy. The teen boys in *Murder by Numbers*, played by Ryan Gosling and Michael Pitt, also quote Nietzsche in their egocentric pursuit of murder. Eventually, whether in the films, or in reality, the detectives close in.

Leopold and Loeb are not the only killers with giant egos who believed they could murder without consequence. Known as the "Dexter Killer" Mark Twitchell spent his life trying to attain fame. His nickname was coined after the popular Showtime series *Dexter* (2006–2013) because of Twitchell's insistence that the titular character's acts (played by Michael C. Hall) inspired him to kill. In *Dexter*, a damaged blood splatter analyst kills by night, exorcising his own demons by murdering people who are criminals themselves.

Although Twitchell, born in 1979, lived far from the glitz of Hollywood, he worked to become a filmmaker in Canada. A huge *Star Wars* fan, he directed an independent prequel to the series, which featured actor Jeremy Bulloch, the original Boba Fett. Twitchell also produced a short horror film and had written a script for a series called *Day Players* (2008). Cracking into the film business is notoriously difficult, so for a man like Twitchell, who was known to have a sizable ego, it began to be too steep of a proverbial hill to climb. He needed a different angle, a revitalization of an old idea. There were thousands of scripts and films about serial killers, but what if he actually became one? What if he documented his real-life murders, never revealing how true they were?

This twisted idea developed into a reality when Mark Twitchell began to pose as a woman on online dating sites. His first victim, Gilles Tetreault, was lured to a garage in Edmonton, believing he was meeting a woman for a first date. Once inside the garage he was brutally attacked by a masked man who prodded Tetreault with a stun baton. Barely conscious and aching from the electric shocks, Tetreault managed to run away from his attacker. Once in safety, though pained and slightly embarrassed, he didn't want to tell the police, or anyone, about his horrific encounter. That is, until Gilles Tetreault learned of the disappearance of Johnny Altinger. Just like Tetreault, Altinger had met who he believed was a woman on PlentyofFish.com and planned to meet her. He told his friends about the plans, and then never came home.

In the meantime, Mark Twitchell was working on his opus, called "SK Confessions" (the SK standing for, of course, serial killer). All written in the first person, this spine-tingling document saved to Twitchell's computer was a moment-by-moment account of not only the attack on Gilles Tetreault but of the murder and subsequent dismemberment of Johnny Altinger's body. It was clear that the budding filmmaker was not going to be satisfied with just one murder and had plans to continue his psychotic need for fame in order to join the pantheon of serial killers.

The "SK Confessions" along with mounting evidence, like Twitchell's inexplicable choice to keep Altinger's car, led to Twitchell's conviction in 2011. He was given a life sentence for this senseless murder and is currently languishing in the Saskatchewan Federal Penitentiary. Despite his status as a prisoner, Mark Twitchell made certain to catch up on all new episodes of *Dexter*, going as far as to buy a TV for his cell.

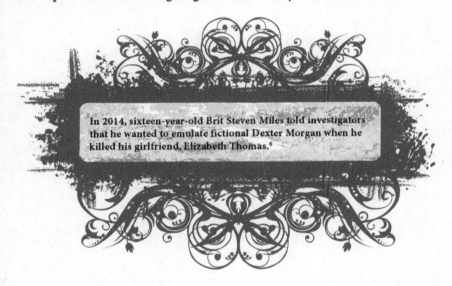

In 2014, sixteen-year-old Brit Steven Miles told investigators that he wanted to emulate fictional Dexter Morgan when he killed his girlfriend, Elizabeth Thomas.[6]

In our previous books we have researched the effects of media on those who perpetrate crimes, and there seems to be little to argue that a TV show or metal song can compel a "normal" human to take a life. There is reason to believe that *Dexter* could have inspired an already troubled, mentally unstable man to go further into his dark fantasy life, making it real. This seems to come up often, like in the case of British serial killer Peter Moore, who killed four people in 1996 at the behest of his imaginary lover, Jason from the *Friday the 13th* (1980) franchise. There was also Pennsylvania teen Michael Anderson who made certain to wear his *A Clockwork Orange* (1971) shirt when he killed Karen Hurwitz with a ninja sword in 1989. His lawyers attempted to use the Stanley Kubrick film's influence as a factor in the killing, but there was simply no proof that a film could be the impetus for such a shocking act, no matter how brutal its scenes. There still isn't proof.

Anderson, like many killers, was found to have psychiatric problems including a personality disorder. Killers like Nathan Leopold, Richard Loeb, and Mark Twitchell are maddening in the misuse of their lives. Though they did not achieve the status of serial killer, they share in the same despicable traits.

CHAPTER EIGHTEEN

Extremely Wicked, Shockingly Evil and Vile (Ted Bundy)

I (Meg) would imagine that most people, when pressed to name a serial murderer, would have Bundy on the tip of their tongue. What is it about Ted Bundy that makes him the quintessential American serial killer? Perhaps it is the disturbing number of victims he stole from this world, or the way in which he played upon their helpful natures with the use of a fake cast on his arm, or his textbook addiction to women with long, brown hair parted in the middle. I would guess it has something to do with his reptilian smile. Many have

Odontologist Richard Souviron explaining bite mark evidence at the Chi Omega trial.

said that Bundy was good-looking and well-mannered. Judge Edward D. Cowart even commented on the murderer's charisma at Bundy's final sentencing. "You're a bright young man. You would have made a good lawyer and I would have loved to have you practice in front of me, but you went another way, partner."[1] This quote was depicted in the Netflix film *Extremely Wicked, Shockingly Evil and Vile* (2019), in which Zac Efron portrayed Bundy. The film, based on the memoir by Bundy's ex-girlfriend Elizabeth Kendall, focuses on Elizabeth's story. We see Ted Bundy through her eyes; charming, loquacious, and a playful father figure to Elizabeth's daughter.

If it is the hidden, twisted shadow beneath the veneer of success that draws us into the world of serial killers, that would explain Ted Bundy's near mythic rise. But allow us to topple his pedestal. Because although he might have appeared clever, Bundy was simply a slave to his own depraved urges. Beneath that sheen of intellectualism, he was nothing more than an empty hull. And no matter how many innocent women he raped, tortured, and killed, he was never satisfied.

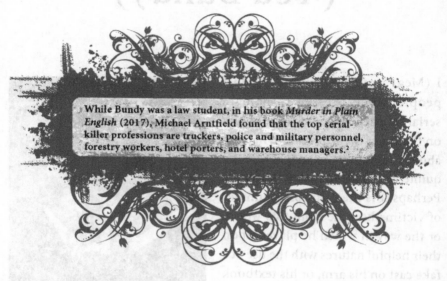

While Bundy was a law student, in his book *Murder in Plain English* (2017), Michael Arntfield found that the top serial-killer professions are truckers, police and military personnel, forestry workers, hotel porters, and warehouse managers.[2]

In *Extremely Wicked, Shockingly Evil and Vile*, we are not privy to much simulated violence. Instead, we see Ted Bundy approach his victims. We see him watch them. We recognize that fake, aforementioned reptilian smile that he used to portray warmth. Ted's unbelievable escapes from the police are shown, as well as his self-important interviews with the press. It is not until the end, when Elizabeth (Lily Collins) has her final confrontation with Ted, that we get a true glimpse of the real man and come to terms with the overwhelmingly violent nature of what he has done.

Throughout this book we explore the psychology behind serial murderers. There is the notion of nature versus nature, the study of personality disorders, as well as the theory of a child's development stunted and then deformed by abuse. But what if there is another environmental factor that we hadn't thought of? Something both so innocuous and so insidious that it was a part of everyday life for centuries. A poison as silent and lethal as Ted Bundy himself that seeped into the air, the soil, and the boards and walls of our houses. This poison, lead, reached an all-time saturation point in America in the 1950s when it was added to gasoline. Every vehicle on the road was soon emitting lead into the environment. At the time, scientists were working feverishly to prove that all that lead in the air, in the paint, *everywhere*, was bad for humans.

It is now known that lead poisoning causes a host of problems from minor headache to agonizing death. And the most terrible aspect of lead's cruel embrace? It affects children the most harshly, as they absorb the toxin more completely in their small, growing bodies. In her published research, Barbara Berney describes a bleak picture; "despite its known toxicity, lead use in the United States increased enormously from the Industrial Revolution through the 1970s, especially after World War II. Between 1940 and 1977, the annual consumption of lead in the United States almost doubled."[3] It wasn't until the 1980s that the scientists were heard and restrictions on gas, paint, and food drastically reduced the lead around us. To compare, blood tests from 1976–1980 to 2015–2016, the geometric mean blood lead level (BLL) of the US population aged one to seventy-four years dropped from 12.8 to 0.82 µg/dL, a decline of 93.6 percent.[4] Today, people who die or grow ill from lead like in the 2014 case of poisoned water in Flint, Michigan, often live in impoverished areas. Lead poisoning affects those with low incomes disproportionately.

Okay, so what does this all have to do with Ted Bundy and his murdering contemporaries? It starts with what we know about how lead affects developing children. Lead was running rampant at a time period in which a shocking amount of serial killers came to be. The 1970s and 1980s are known as the serial killer "heyday," think Bundy, Dahmer, Gacy, and many more. In fact, serial killing has plunged a monumental 85 percent in three decades! Many have theorized that this rise several decades ago has to do with the popularity of hitchhiking, as well as a more relaxed style of parenting, in which kids were often alone. Others have a more scientific theory. They believe that the preponderance of lead in the environment caused not only a spike in serial killing, but in all aggressive crime.

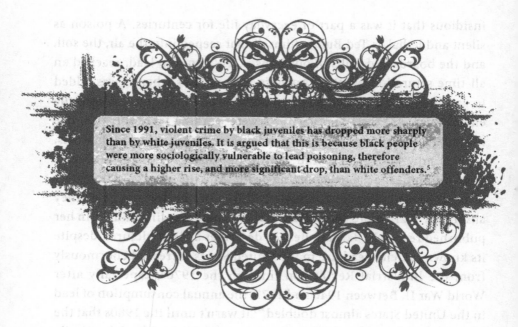

Since 1991, violent crime by black juveniles has dropped more sharply than by white juveniles. It is argued that this is because black people were more sociologically vulnerable to lead poisoning, therefore causing a higher rise, and more significant drop, than white offenders.[5]

When I first heard about the lead murder theory, I was skeptical. Could a toxin in the air really exacerbate or even cause someone to kill? When pediatrician Herb Needleman began his practice in the 1950s, he became acquainted with the serious blight on public health that lead had become. PBS.org reports:

> Throughout the 1960s, '70s, and '80s, Needleman crusaded for more stringent lead safety standards after his groundbreaking research showed that chronic, low-level lead poisoning could cause devastating neurological impairments in children. Before the publication of his work, researchers believed that lead poisoning was a short-term problem that could easily be treated. Yet his work showed that the effects could be lingering and devastating—even if a child didn't show overt signs of lead poisoning.[6]

"Neurological impairments" is a rather broad term. As we dug deeper, we found that aggression was included under this umbrella. In a study conducted in St. Louis, Missouri, in 2017, scientists found a significant connection between blood lead levels and acts of aggressive crime, including armed robbery and murder. Even with the inclusion of important sociological factors, this connection persisted. There is a plethora of studies over the decades with the same findings. Lead in the brain remains there, and it has a link to violent crime.

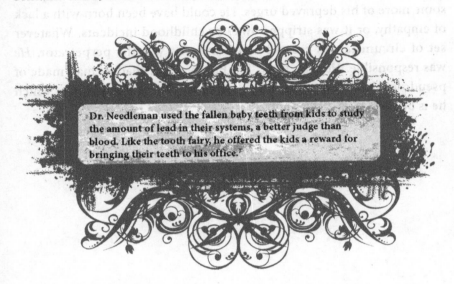

Dr. Needleman used the fallen baby teeth from kids to study the amount of lead in their systems, a better judge than blood. Like the tooth fairy, he offered the kids a reward for bringing their teeth to his office.

The rate of aggressive crime (like serial murder) rose starkly in the adult years of those exposed to the most lead. Since the decline of lead, there is a striking decline in crime. We've seen this same rate of decline in other countries that banned lead. Jamaica banned lead in gasoline in the 1990s and 2000s and currently has a drastic drop in violent crime to show for it.

Journalist Kevin Drum has long been researching and uncovering this phenomenon. Drum argues that there is no coincidence between these rises and falls. He explains that there have been two "epidemics of murder" in the US—one in the 1920s–1930s and the other in the 1970s–1980s—and that they both can be linked to lead spikes in our environment. While it is a fascinating theory, to think that Ted Bundy was exposed to lead in his childhood and that it changed his brain chemistry, Kevin Drum makes clear that the substance does not create a monster from nothing. "It's important to emphasize that the lead-crime hypothesis doesn't claim that lead is solely responsible for crime. It primarily explains only one thing: the huge rise in crime of the seventies and eighties and the equally huge—and completely unexpected—decline in crime of the nineties and aughts."[8]

Perhaps that leaves us where we started. Lead could explain some added aggression, the pornography Bundy watched could have inspired some more of his depraved urges. He could have been born with a lack of empathy, or it was stripped away by childhood incidents. Whatever set of circumstances created Ted Bundy, *he* was the perpetrator. *He* was responsible. *He* took lives. Ted Bundy was a shadow man made of pseudo-charm. He deserves no pedestal, only the dark soil under which he is buried.

SECTION SEVEN
TRAVELING KILLERS

CHAPTER NINETEEN

American Crime Story: The Assassination of Gianni Versace (Andrew Cunanan)

American Crime Story: The Assassination of Gianni Versace (2018) brings the viewer into the world of murderer Andrew Cunanan, who killed not only Gianni Versace but also four others in the spring and summer of 1997. Told non-linearly, it's a fascinating look inside the mind of Cunanan and the lives of his victims. Writer Tom Rob Smith said:

> It was never going to be a story told through the point of view of the police, which is normally when it becomes pure crime drama. Part of this is because there was never a central piece to the killing of Versace. You have the police investigating in Minneapolis, police investigating in Chicago, police investigating in Miami, they never really linked up. There was never a cohesive story from a traditional crime point of view."[1]

This interesting perspective makes for a thrilling nine-episode series.

Andrew Cunanan's victims included Jeffrey Trail and David Madson in Minnesota, Lee Miglin in Illinois, William Reese in New Jersey, and famous designer Gianni Versace in Florida. Cunanan was known as a compulsive liar. He would tell tall tales and was able to change his appearance according to what he felt was most attractive at a given moment.[2] What is the science of lying? Studies show that lying rewires your brain. The more you lie, the better you become at it.

In a 2016 study in the journal *Nature Neuroscience*, [Duke psychologist Dan] Ariely and colleagues showed how dishonesty alters people's brains, making it easier to tell lies in the future. When people uttered a falsehood, the scientists noticed a burst of activity in their amygdala. The amygdala is a crucial part of the brain that produces fear, anxiety, and emotional responses, including that sinking, guilty feeling you

Serial killer Andrew Cunanan.

get when you lie. But when scientists had their subjects play a game in which they won money by deceiving their partner, they noticed the negative signals from the amygdala began to decrease. Not only that, but when people faced no consequences for dishonesty, their falsehoods tended to get even more sensational.[3]

We all lie. We may lie to protect other people's feelings, called a pro-social deception, or to hurt others, called anti-social deception. The intention behind the lie matters. Children are often taught to lie at a young age by being told to pretend to like a gift or food, even if they don't, to appear polite. Whether lying becomes a problem is based on the lie's objective.

Cunanan was also known to have an addiction to painkillers. How could this have affected him? Painkillers can cause harmful effects on the brain and body. "Painkillers cause chemical changes to the brain and also kill brain cells. The most affected areas of the brain are those areas that deal with cognition, learning, and memory."[4] Cunanan was also a heavy drinker. Combining painkillers and alcohol is very dangerous and could cause nausea and vomiting, changes in blood pressure, marked disinhibition, abnormal behavior, and possible coma.[5]

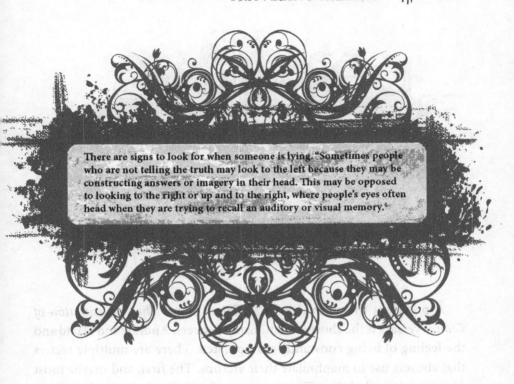

There are signs to look for when someone is lying. "Sometimes people who are not telling the truth may look to the left because they may be constructing answers or imagery in their head. This may be opposed to looking to the right or up and to the right, where people's eyes often head when they are trying to recall an auditory or visual memory.[6]

In the episode "House by the Lake," Andrew (Darren Criss) ties up David's (Cody Fern) dog before the murder of Jeffrey (Finn Wittrock). How do dogs react to violence? Dogs are known as intuitive creatures and are often seen trying to break up fights or distract quarreling owners. "Science has confirmed that dogs can feel primary emotions such as sadness, fear, or joy. Although it is not confirmed from a scientific stand-point, many researchers and dog owners alike agree that it is possible that dogs can feel and experience secondary emotions as well."[7] How would a dog react differently to a stranger or intruder though? In a study conducted in 2017, two men dressed in protective bite suits entered the homes of three separate volunteers who owned dogs. All three owners were confident that their dogs would attack. In the first case, the dog barked but stayed far away from the men. In the second case, the dog wagged its tail and even let the men steal some items from the home. In the third, the dog barked but did not attack.[8]

According to a study from the CDC, approximately 4.5 million dog bites occur in the United States each year, and 800,000 of those bites result in medical care.[9]

A prevalent theme in *American Crime Story: The Assassination of Gianni Versace* is the abusive relationship between Andrew and David and the feeling of being constantly manipulated. There are multiple tactics that abusers use to manipulate their victims. The first, and maybe most common, is gaslighting. This is a type of emotional abuse wherein an abuser lies or downplays the impact of an event or something that they say. Abusers can also use isolation tactics to control who their victim talks to or what media they consume. Abuse may show up in the form of overprotection which is disguised as an abuser being worried about the victim and their safety. Emotional manipulation is often used, especially when swinging between anger and apologies. The victim may be made to feel guilty for feeling scared or sad. Threats and physical abuse often occur. "On average, more than one in three women and one in four men in the US will experience rape, physical violence, and/or stalking by an intimate partner."[10]

No one chooses to get into an abusive relationship. When I (Kelly) first heard the analogy of a frog in a pot of boiling water, I was in therapy working through the remaining trauma of an abusive relationship from my own past. The story is that if a frog jumped into a pot of boiling water, it would immediately leap out. If a relationship started out as abusive, very few people would choose to stay. But, if a frog jumped into a pot of warm water that was slowly brought up to a boil, by the time the water was boiling it would be too late to jump out. This is like abusive relationships. By the time you realize you're in one, it feels too late to get out. Thankfully, my situation didn't end in tragedy, but it's important to be aware of your options for getting help and being safe.

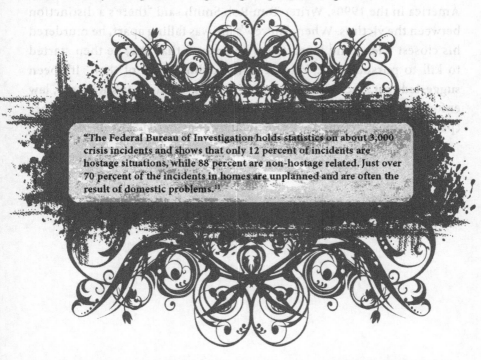

"The Federal Bureau of Investigation holds statistics on about 3,000 crisis incidents and shows that only 12 percent of incidents are hostage situations, while 88 percent are non-hostage related. Just over 70 percent of the incidents in homes are unplanned and are often the result of domestic problems.[11]

After the murder of Jeffrey, David is basically being held hostage by Andrew. What is the mentality that victims experience during these types of situations?

Hostage situations can generally be described in two basic ways. One is the traditional situation, in which the hostage-taker tries to utilize hostages as leverage to negotiate something else. The other situation, which is becoming more prevalent, occurs when the hostage-taker is bent on death and destruction to 'make a statement' and has no other goal in mind. As a situation develops, a potential hostage needs to immediately assess the intruder's intent; negotiation or murder."[12]

Hostage victims may suffer from long lasting effects of their experience including impaired memory and concentration, shock and numbness, and social withdrawal.

As Cunanan was an out gay man, the series explores homophobia in America in the 1990s. Writer Tom Rob Smith said "there's a distinction between the victims. When Andrew's life was falling apart, he murdered his closest friend and lover. Once he crossed that line, he then started to kill to pursue ideas. Versace is the culmination of that." It's been suggested that since the first several victims of Cunanan's were gay, law enforcement may not have responded as seriously. Like Jack the Ripper (page 9), some victims are unfortunately considered "less dead."

CHAPTER TWENTY
Henry: Portrait of a Serial Killer (Henry Lee Lucas)

To prepare for his audition for *Henry: Portrait of a Serial Killer*, Michael Rooker watched an interview between Barbara Walters and Henry Lee Lucas. Michael comments on this interview:

> I thought he was full of shit when I was watching it. He was just saying all sorts of stuff, and you could tell he just got off on that, and I didn't know if I believed even half of what he was saying. But I ended up taking away a little bit of body language and some of his vocal rhythms. That helped a lot, certainly a lot more than any of the books that were being written about psychopaths and sociopaths at the time.[1]

Lucas was considered a fabulist and gave many false confessions. What is the science behind this condition? This, like Andrew Cunanan in the last chapter, is related to compulsive lying. Robert Reich, MD, a New York City psychiatrist and expert in psychopathology, says "compulsive lying has no official diagnosis. Instead, intentional dissimulation, not the kind associated with dementia or brain injury, is associated with a range of diagnoses, such as antisocial, borderline, and narcissistic personality disorders."[2] Some experts think

Henry Lee Lucas was granted TV privileges, milkshakes, and fancy steak dinners by law enforcement in exchange for confessions, which turned out to be false.[3]

that the advent of social media is enabling fabulists and compulsive liars. There are other reasons, though, why someone may confess to a crime they did not commit. There have been over three hundred and fifty wrongful convictions overturned by DNA evidence and many of them involved a false confession. Reasons for these false confessions include the real or perceived intimidation of the suspect by law enforcement, a compromised reasoning ability of the suspect, devious interrogation techniques, or fear, on the part of the suspect, that failure to confess will yield a harsher punishment.[4]

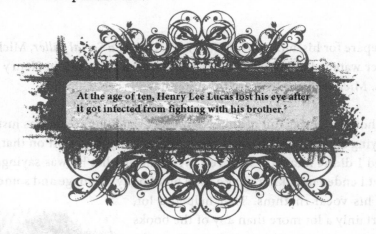

At the age of ten, Henry Lee Lucas lost his eye after it got infected from fighting with his brother.[5]

Lucas was given access to police records to "refresh" his memory. What are the ethics behind this practice? As Louis Pastuer said, "in the field of observation, chance favors only the prepared mind."[6] Lucas's confessions, which occurred before the advent of DNA technology, tied him to close to six hundred murders. Most of these would prove to be false, according to timeline analysis, but his access to records helped enhance his credibility. His confessions helped give false closure to numerous families of murder victims. "Lucas's story is a testament to the disaster that can follow when police seek easy answers, in this case taking the claims of a con man as a quick way out of stalled detective work."[7] Journalists began to uncover the unrealistic timelines before law enforcement did. This is reiterated by private investigator Steven Mason:

A well-developed timeline provides clarity to a complex case, and it should be used as a reference to understand important relationships. By studying a timeline, both the investigator and defense counsel can find gaps in the investigation, document the movements of witnesses and victims, exploit inconsistencies, develop alibis, and evaluate the plausibility of the opposing counsel's case.[8]

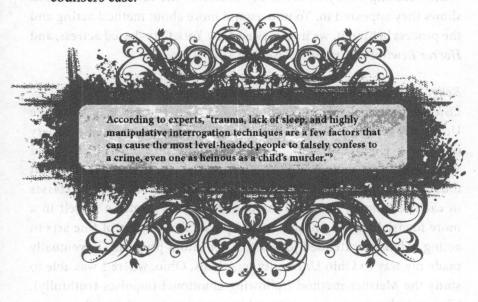

According to experts, "trauma, lack of sleep, and highly manipulative interrogation techniques are a few factors that can cause the most level-headed people to falsely confess to a crime, even one as heinous as a child's murder."[9]

We can't talk about Henry Lee Lucas without mentioning his connection with Ottis Toole. He was a drifter and serial killer who was convicted of six counts of murder.

Like his companion Henry Lee Lucas, Toole made confessions he then later recanted, which resulted in murder convictions. The discrediting of the case against Lucas for crimes for which Toole had offered corroborating statements created doubts as to whether either was a genuine serial killer or, as Hugh Aynesworth suggested, both were merely compliant interviewees whom police used to clear unsolved murders from the books.[10] An interesting connection between the two is that they both had troubled childhoods which included being dressed up as girls, subjected to humiliation, and other unpleasantries. Toole was portrayed by Tom Towles in the film and died in 1996 in prison from cirrhosis.

Something I (Kelly) have noticed during research for this book is that many actors chose to pursue their roles as serial killers by using method acting. Michael Rooker reportedly showed up "in character" every day while shooting *Henry* as did many others in the movies or television shows they appeared in. To understand more about method acting and the process behind it, we interviewed New York City–based actress, and *Horror Rewind* contributor, Lisa Bol:

Kelly: "Tell us a bit about your training."
Lisa Bol: "I received my bachelor of arts in theatre arts from the University of Minnesota, Twin Cities. While there I studied a variety of styles: LeCoq (clowning), Margolis Method (movement-based), and Stanislavski (psychologically/emotionally-based). It was a really great foundation to start from, but the program only offered a couple classes in each technique. I wanted the opportunity to immerse myself in a more focused training program, so I pursued a masters of fine arts in acting. After an extended journey to get into a program, I eventually made my way to Ohio University in Athens, Ohio, where I was able to study the Meisner method (following emotional impulses truthfully), Chekov technique (movement- and imagination-based), and those were supplemented with voice, speech, movement, and additional classes to prepare a professional actor. Following my MFA from Ohio University, I spent a year as an acting apprentice at the Actors Theatre of Louisville where I was able to put my training to work as well as watch and learn from working professionals."

Meg: "How do you prepare for a role? How do you get into the head of a character?"
Lisa Bol: "Research, research, research! After reading the script a couple of times, I look into the world of the piece—when and where it is taking place and how that differs from what I know to be true personally. If it's set in a place I've never been, I will research that location—look at pictures to help me visualize the world. If it's in a different time period,

then I want to be familiar with things like clothing (that affects movement and much more), customs/sensibilities (what women were allowed to do and how were they 'expected' to behave), and what else was going on in the world at that time (war, hardships, prosperity.) If I'm working on a play and there are either films of it or I can find archived footage of it, I love watching what others have done before me—mostly to help me determine what I feel doesn't work and to see if I can find any additional insight into the world and the character. If I'm preparing for a television or film role, I will watch episodes from the series or something from the genre to help me understand the tone of the world.

As I'm getting a hold of the when and the where, I'm also beginning to get a handle on the 'who,' who is my character (age, life experiences, dreams, etc.) and who are the other characters in the piece. I then track how my character interacts with the other characters in the piece and what my relationships are to each of them and how I feel about them. Which then of course leads us to 'why,' why my character makes the choices they do. It may seem like a lot, but it is definitely a process that becomes routine and second nature."

Meg: "What are the potential dangers of method acting or getting too close to a character?"
Lisa Bol: "There are so many potential dangers of method acting: coming into physical harm because you want to really be in the fight as opposed to using stage combat or choreography, addiction or death from experimenting with illegal substances just so you know what it feels like to have a specific high, damage to personal relationships because you're actively being someone else, and even carrying the trauma of the character into your own life."

Kelly: "How do you prepare to play a character who is vastly different from yourself?"
Lisa Bol: "More research! Finding the small things that are similar between us and then building on top of that. What adversities have we both overcome? Have we experienced love in the same way? Do we share hobbies? I ask questions like that and when I stop seeing the similarities, I start to look at the differences and ask myself what got them to where

they are compared to where I am in life. Often lots of imagination comes into play here to fill in those gaps because it isn't always in the text. Once I feel I understand why they are the way they are, then it is my job to tell their story truthfully and without judgement. For me, playing a character who is vastly different from myself is sometimes the most fun—even when it gets dark and challenging."

Kelly: "What do you want people to know about the craft of acting?"
Lisa: "Very often, a lot of care and work goes into the creation of a character—from the writer's end as well as the actor's. My fellow actors and I are always amused when an audience member asks us 'how did you memorize all those lines?' as though that was the most difficult part of the process. While memorization isn't always easy, it's the work behind the words that's the real challenge."

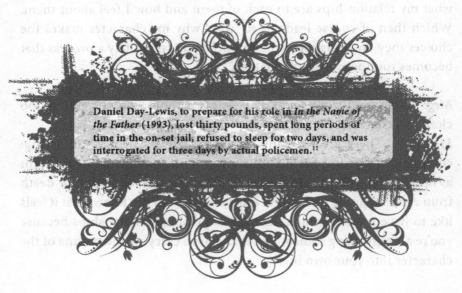

Daniel Day-Lewis, to prepare for his role in *In the Name of the Father* (1993), lost thirty pounds, spent long periods of time in the on-set jail, refused to sleep for two days, and was interrogated for three days by actual policemen.[11]

Some acting instructors have taken to teaching a scientific approach to method acting. David Ihrig, a drama lecturer at University of California-Irvine, uses a technique that focuses on cognitive and behavioral science. This infuses onstage characters with genuine memories, feelings, and personalities, bringing a new level of realism to the performing arts.[12]

A neurologist who frequently lectures, James McGaugh, says that if he were an actor, he would create an autobiography for his character, generate memories and a personality for his character, and then give the character a rich world to live in. This cross-disciplinary teaching technique can benefit both scientists and artists in a safe and monitored way. Regardless of the acting methods used in this film, *Henry* offers a chilling and unsettling portrayal of a real-life serial killer.

CHAPTER TWENTY-ONE
American Predator
(Israel Keyes)

The disappearance of a young woman named Samantha Koenig in Anchorage, Alaska, sparked a nationwide manhunt that would lead to the discovery of one of America's most notorious serial killers: Israel Keyes. He admitted to killing less than a dozen people; the FBI believes the number to be eleven, but only three of his victims were definitively identified: Koenig in Alaska and a couple, Bill and Lorraine Curriers, in Vermont. Keyes claimed he took at least five other lives but never named these victims.

Israel Keyes is a rare serial killer in that he did not have a victim profile.

Per his account, he killed four people in Washington state: a couple sometime between 2001 and 2005, and two separate victims in 2005 and 2006. Keyes also stated that in 2009 he murdered someone on the East Coast, then left the body in New York state. The FBI is "relatively confident" that this victim was Debra Feldman, a New Jersey resident who went missing in April 2009.[1]

Keyes was homeschooled as a child. There is a lot of research that focuses on the effects of socialization in children who are homeschooled. Most of this research finds that "being homeschooled does not harm children's development of social skills. In fact, some research finds that homeschooled children score more highly than children who attend school on measurements of socialization."[2] The only problem with this research is that it relies on volunteer participants and is entirely self-reported. Opponents to homeschooling suggest that socialization isn't just about seeing other people and having conversations.

Socialization requires that children consistently work with people they're not used to working with. It's about discussing things with people who have a different opinion and challenging preconceived notions. It's about having to do a group project with people who don't necessarily work the same way as you do, to collaborate on ideas and grow as a thinker.[3]

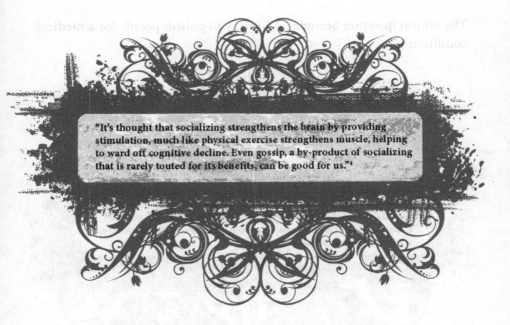

"It's thought that socializing strengthens the brain by providing stimulation, much like physical exercise strengthens muscle, helping to ward off cognitive decline. Even gossip, a by-product of socializing that is rarely touted for its benefits, can be good for us."[4]

The fundamentalist religious background of his youth could have had a massive effect on Keyes's ideologies. "The domain of ideology is of unique interest because it is hypothesized to contain elements of both facts and preferences."[5] Being inundated with a racist, anti-Semitic, and white supremacist interpretation of Christianity most likely affected Keyes's view of the world and those around him.

Keyes set fires at some of his crime scenes. Arson can be motivated by a desire to destroy but can also be an irresistible compulsion. Fire setting is recognized as a form of mental illness and usually stems from another deep-seated pathology.

Research shows that fire setters are significantly more likely to have been registered with psychiatric services compared with other criminal offenders. Between 10 to 50 percent of patients who are admitted to medium-security forensic mental health services have a record of deliberate fire setting. Fire setting in adolescence and early adulthood predicts schizophrenia in later life. Fire-setting behavior is associated with animal cruelty in juveniles; the other statistically significant risk factors being male gender, and the victim of sexual abuse."[6]

The ethical question becomes: Is it okay to punish people for a medical condition?

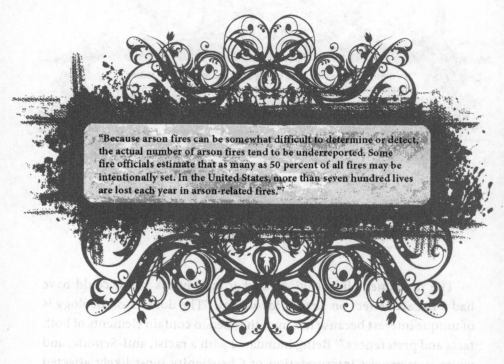

"Because arson fires can be somewhat difficult to determine or detect, the actual number of arson fires tend to be underreported. Some fire officials estimate that as many as 50 percent of all fires may be intentionally set. In the United States, more than seven hundred lives are lost each year in arson-related fires."[7]

Surveillance footage from the night of Koenig's abduction shows a man leading her away from her workplace into a white truck. How often is surveillance video used to catch perpetrators? Video surveillance technology has been around since 1942 and has progressed in quality and availability. Since so much footage can be kept longer now, due to the smaller size of memory cards, police are able to access databases of past footage to connect criminals to crimes. The availability of surveillance cameras for the general public has also proven to help catch culprits either in the act of a crime or after the fact.

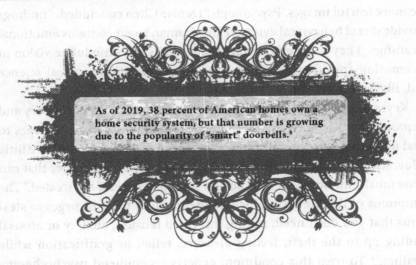

As of 2019, 38 percent of American homes own a home security system, but that number is growing due to the popularity of "smart" doorbells.[8]

Typically, the FBI will be able to come up with a profile for a potential abductor or killer. They can figure out their modus operandi, scour social media and phone records for clues, and track potential suspects' whereabouts. In the Koenig case, the picture wasn't coming together. Part of the reason for this was because Keyes planned things out meticulously ahead of time. On his murder trips, he kept his mobile phone turned off and paid for items with cash. He had no connection to any of his victims.

Keyes also left "murder kits" in various locations around the country that contained, among other items, weapons and cash. The money came from bank robberies he committed to support his criminal activities. The caches provided further cover because Keyes didn't have to risk boarding an airplane with a weapon or using credit cards that could later connect him to a crime in a particular area.[9]

These murder kits were considered a signature of his.

Maureen Callahan, author of *American Predator* (2019), said "when Keyes took people, he was acutely attuned to their animal response: the acid flush of adrenaline flooding the brain, color draining from the faces, pupils dilating in fear. He could smell it in their sweat. He liked to extend that response as long as possible."[10] Is it possible to smell fear in someone's sweat? In one study, samples were collected of fearful sweat from male volunteers who watched films that incited fear. Female volunteers then viewed images while smelling the samples and were found to be biased to the more fearful images. Psychologist Denise Chen concluded, "findings provide direct behavioral evidence that human sweat contains emotional meanings. They also demonstrate that social smells modulate vision in an emotion-specific way."[11] Olfactory communication is a real science and, like emotions, it can be contagious.

Keyes was a notorious robber. What is the psychology of robbery and kleptomania? Kleptomania is "the recurrent inability to resist urges to steal items that you generally don't really need and that usually have little value. Kleptomania is a rare but serious mental health disorder that can cause much emotional pain to you and your loved ones if not treated." The symptoms of this condition include the inability to resist urges to steal items that you don't need, feeling increased tension, anxiety or arousal leading up to the theft, feeling pleasure, relief, or gratification while stealing.[12] To treat this condition, experts recommend psychotherapy for impulse control and medication to curb those impulses.

If Keyes had been such a careful, meticulous planner, how was it that he got caught for the murder of Samantha Koenig? He told law enforcement that he'd been feeling out of control and noted, "back when I was smart, I would let them come to me."[13] He broke one of his own rules by killing close to home and gave law enforcement plenty of clues, like using Koenig's ATM card in Texas, that led to his arrest. Keyes died by suicide three months before his scheduled trial. We may never know the extent of Keyes's crimes, but one thing is certain: he truly was a predator that hunted for sport.

SECTION EIGHT
NEVER CAUGHT

CHAPTER TWENTY-TWO

Zodiac
(The Zodiac Killer)

Just because a serial killer has never been apprehended by law enforcement doesn't mean that they can't become a celebrity of mythic proportions. Like Jack the Ripper, the Zodiac Killer has become a looming shadow, casting his darkness on the media wide and far. The most famous iteration is *Zodiac* (2007), directed by David Fincher and starring Jake Gyllenhaal who portrays real life Robert Graysmith, a political cartoonist who became fixated on solving the Zodiac Killer's cryptograms. There are numerous films and TV series where the reign of the Zodiac has made its indelible mark, from 1971's *Dirty Harry* in which Clint Eastwood's titular character hunts serial killer "Scorpio" to the recent series *Riverdale* (2017–) in which a mysterious killer named the "Black Hood" sends ciphers to Betty (Lili Reinhart) and the gang.

It is no wonder that the Zodiac Killer looms large in our collective consciousness. He was a sadistic killer who targeted kids and young adults at their most vulnerable, sharing intimate moments on Lover's Lane. Responsible for at least seven murders and two attempted murders, the Zodiac would do well in the "I want to be famous" section of this book, as he clearly enjoyed the spotlight. The only twist is, fifty years later the Zodiac is still only an empty mask. Beneath the Hollywood treatment, and the real victims' harrowing attacks, there is an invisible man.

The Zodiac is certainly not the first serial killer to recede into the night, free from prosecution. What is unique in his case, is his frequent communication with the press and, subsequently, the authorities. These letters were filled with threatening, and often poorly spelled, phrases. For example, on April 20, 1970, a letter arrived with a drawn diagram of a bomb that Zodiac claimed he was going to use to blow up schoolchildren. Though, in the same letter he explained that "there is more glory in

Some believe that the Phantom Killer of Texarkana and the Zodiac Killer are one and the same. The Phantom Killer also killed young people on Lover's Lane, used the same caliber of firearm, and wore a similar masked disguise. The Phantom Killer is memorialized in the film *The Town That Dreaded Sundown* (1976)[1]

killing a cop than a cid (sic) because a cop can shoot back."[2] It seems his killing philosophy was rather confused. Many of his letters were sent to the *San Francisco Chronicle* and other news sources because Zodiac liked the notion of his words being printed for all of Northern California, and conceivably the world, to see. His way of assuring this was to threaten more violence in his stilted language: "If you do not print this cipher by the afternoon of Fry. 1st of Aug 69, I will go on a kill rampage Fry. night. I will cruise around all weekend killing lone people in the night then move on to kill again, until I end up with a dozen people over the weekend."[3]

The newspaper editor, with the cooperation of the police, published the cipher given to them in the letter on the fourth page of the *Chronicle*, asking for help in solving the coded string of symbols which included pyramids, slashes, and circles. They took the threat seriously, as the sender had taken responsibility for the shooting deaths of teenagers Betty Lou Jensen and David Faraday on Lake Herman Road in December of 1968, as well as the murder of twenty-two-year-old Darlene Ferrin, and the attempted slaying of Michael Mageau on July 4, 1969 at Blue Rock Springs Park in Vallejo. What really captured the attention of the public was the cipher itself, a puzzle to be solved that would hopefully allow the police to apprehend this murderer who was brazenly challenging them to unmask his identity.

The 408 Cipher is considered a homophonic cipher. The easiest way to break this code is to look for frequencies of symbols and match them with common phrases and often used letters like E.[4]

To understand ciphers, one should first become acquainted with cryptography itself. The true science and practice of cryptography is rooted in communication that can be used to prevent third parties, or adversaries, from deciphering the meaning. Even before modern technology, there were many kinds of ciphers used to obfuscate important messages. The Caesar Cipher, named for Julius Caesar, was used by the emperor in his military letters. It is a rather simplistic cipher, where the placement of a letter is shifted. Specifically, Emperor Caesar shifted A four spots to D, effectively shifting the entire alphabet four letters. Therefore, B would be E, C would be F, etc. This use of the cipher was corroborated by Roman historian Suetonius. "If (Caesar) had anything confidential to say, he wrote it in cipher, that is, by so changing the order of the letters of the alphabet, that not a word could be made out."[5]

Edgar Allan Poe wrote a winning short story about cryptography in 1843. He earned a hundred-dollar prize, the most he ever won, for "The Gold-Bug," about men deciphering a cipher to figure out their way to a buried treasure. At the time of its publication, cryptography and secret writing was of avid interest to Victorians. Author Simon Singh explains why ciphers and secret messages could prove useful in the rather prudish era:

> As people became comfortable with encipherment, they began to express their cryptographic skills in a variety of ways. For example, young lovers in Victorian England were often forbidden from publicly expressing their affection and could not even communicate by letter in case their parents intercepted and read the contents. This resulted in lovers sending encrypted messages to each other via the personal columns of newspapers. These "agony columns," as they became known, provoked the curiosity of cryptanalysts, who would scan the notes and try to decipher the titillating contents. Charles Babbage is known to have indulged in this activity, along with his friends Sir Charles Wheatstone and Baron Lyon Playfair, who together were responsible for developing the deft Playfair cipher. On one occasion, Wheatstone deciphered a note in *The Times* from an Oxford student, suggesting to his true love that they elope. A few days later, Wheatstone inserted his own message, encrypted in the same cipher, advising the couple against this rebellious and rash action. Shortly afterward there appeared a third message, this time unencrypted and from the lady in question: "Dear Charlie, Write no more. Our cipher is discovered."[6]

Cryptography made a significant impact on World War II, in which codebreakers were needed to decipher both Japanese and German correspondence. The German code machine Enigma was considered unbreakable, that is until mathematician Alan M. Turing and his team were able to intercept Nazi naval plans. It is believed that their mathematical and statistical genius were indispensable in defeating the Nazis by circumventing their naval attacks. Turing's story came to the big screen in 2014's *The Imitation Game* starring Benedict Cumberbatch as the mathematician.

No one knows how the Zodiac Killer came to be interested in ciphers, perhaps from these examples in history, or as a passing hobby. The first

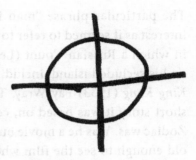

of the Zodiac's four ciphers was sent in
three parts to three separate newspapers.
Decades later it is the only one that has
been fully solved. Considered to be
the simplest, it took a North Salinas,
California, couple only a week to decipher. Donald Harden, a schoolteacher,
and his wife Bettye were intrigued by the
puzzle. They worked on making sense of
the "Zodiac Alphabet."

The Zodiac Killer used this
cross-circle symbol in his letters.

Bettye is credited with discovering two cribs, words or phrases suspected to appear in the message. Cribs are powerful cryptanalytic
tools, because once a location or locations can be determined for
them, several substitutions can be identified, which can accelerate
the unravelling process. Inspired by the killer's obvious craving
for attention, Bettye guessed that the message would begin with
the word "I." She also believed the word KILL or KILLING—or
even the phrase I LIKE KILLING—would appear somewhere in
the message. Her guesses turned out to be correct.[7]

It's fascinating to think that it was not only Bettye Harden's logic, but also
her understanding of what kind of man the Zodiac might be, that truly
unlocked the code. The FBI concurred with the Hardens. The solved
code reads as follows:

I like killing people because it is so much fun—it is more fun
than killing wild game in the forest because man is the most
dangerous animal of all—to kill something gives me the most
thrilling experience—it is even better than getting your rocks
off with a girl—the best part of it is that when I die I will be
reborn in paradise and all the (lone or stray people) I have killed
will become my slaves—I will not give you my name because
you will try to slow down or stop my collecting of slaves for
my afterlife.[8]

The particular phrase "man is most dangerous game of all" piqued interest as it seemed to refer to the 1932 film *The Most Dangerous Game* in which a Russian Count (Leslie Banks) hunts unsuspecting humans on his secluded island, including Eve Trowbridge, played by the star of *King Kong* (1933) Fay Wray. This film reference, or perhaps the 1924 short story it was based on, could be considered a clue as to who the Zodiac was. Was he a movie buff who frequented theaters? Had he been old enough to see the film when it was originally released? Did he see killing as sport?

There was also a string of eighteen letters at the end of the cipher that was never solved. Some have speculated that these scrambled letters may be the Zodiac's name made with a different kind of more challenging code.

The Zodiac sent his three remaining ciphers over the next year. They did not use the same simplistic alphabet-swap style as the first, leaving people confused by how to solve them. Even modern codebreaking software struggled to determine the cipher.

In a coincidental twist of fate, less than a week after finishing the first draft of this chapter, the world awoke to news about the Zodiac. After fifty-one years, the second cipher had been solved! It reads as follows:

I hope you are having lots of fun in trying to catch me
That wasn't me on the TV show which brings up a point about me
I am not afraid of the gas chamber because it will send me to paradice (sic) all the sooner Because I now have enough slaves to work for me where everyone else has nothing when they reach paradice so they are afraid of death
I am not afraid because I know that my new life will be an easy one in paradice death.[9]

The TV show referenced is *AM San Francisco* in which a person called claiming to be the Zodiac and spoke to host Jim Dunbar. It was a brief call that led to no new clues. This newly solved cipher is similar in that it doesn't give a lot of clues as to the identity of the Zodiac, but it does mean that his remaining ciphers are most likely solvable, and they may hold the key to who he is. Known as the 340 cipher, it took David Oranchak, a software developer in Virginia, Jarl Van Eycke, a Belgian computer programmer, and Sam Blake, an Australian mathematician, nearly fifteen years to solve. Oranchak described how it felt to finally find the solution. "It was incredible. It was a big shock, I never really thought we'd find anything because I had grown so used to failure. When I first started, I used to get excited when I would see some words come through—they were like false positives, phantoms. I had grown used to that. It was a long shot—we didn't even really know if there was a message."[10]

The Zodiac Killer has not only inspired many films, books, and TV series, it has also brought about a flurry of theories and accusations. Some believe the Unabomber Ted Kaczynksi could've been responsible, as he lived in Northern California at the time of the killings and had a similar interest in cryptography. After an FBI investigation in 1996, authorities were satisfied that he was not the Zodiac.

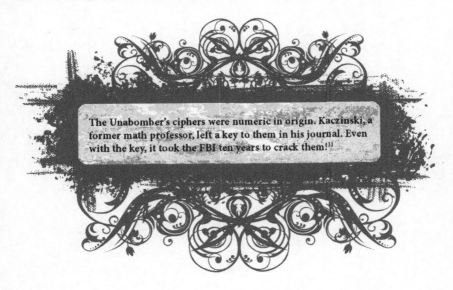

The Unabomber's ciphers were numeric in origin. Kaczinski, a former math professor, left a key to them in his journal. Even with the key, it took the FBI ten years to crack them![11]

In 2002, Randy Kenney from upstate New York went to the San Francisco police with a fascinating tale. He admitted that his best friend, Louis Myers, confessed that he was the Zodiac Killer. Myers asked Kenney not to reveal this until after his death. There have been many false confessions, especially for such a media-fueled case, but there seems to be some evidence that Myers could be the killer. Myers went to high school with the first two victims and worked at the same restaurant with another. He was also stationed overseas with the military during the same years that the Zodiac was inactive. While these coincidences, as well as a confession, pique curiosity, there is just not enough to prove that Louis Myers was guilty. There have even been recent rumblings that Donald Harden, the man who solved the first cipher with his wife, could be a likely candidate. There is hope with the genealogical DNA advancements that were used to catch the Golden State Killer, Joseph DeAngelo, in 2018, that the Zodiac will not always be a killer hiding behind strings of code.

Seven young people lost their lives because of the Zodiac Killer. It's easy to be intrigued by his strange letters and perhaps even admirable ciphers, but we can't fail to recognize his legacy of death and suffering.

CHAPTER TWENTY-THREE

Wind River
(The Highway of Tears)

Serial killers are as real as any natural disaster that gorges itself on destruction, ripping apart everything you have ever loved. What's worse is that while there are serial killers with notorious names, there are just as many without. Like the Servant Girl Annihilator of Austin, Texas, who murdered eight women in the mid-1880s. As his moniker suggests, he killed women of the servant class, most of them Black. They had all been asleep in their beds. This faceless monster disappeared into the fabric of America, just like the Freeway Phantom of 1970s Washington D.C., who killed at least six Black teens and women before vanishing.

The Freeway Phantom left a note in the pocket of murder victim Brenda Woodard's jeans. It dared the police "to catch me if you can." Because DNA evidence was lost in the case, the Phantom may never be caught.[1]

One ephemeral killer (or a few, as some insist) has situated himself on the Yellowhead 16 Highway in Northern British Columbia. This rural stretch is characterized by poverty. With no public transportation until recently, many had to hitchhike on the heavily forested road, the perfect scenario for a serial killer in their midst.

The exact number of women who have disappeared or been murdered along Highway 16 is disputed. The Royal Canadian Mounted Police acknowledges eighteen murders and disappearances in its list of Highway of Tears cases, dating

The Highway of Tears is a 450-mile corridor of Highway 16 between Prince George and Prince Rupert, British Columbia, Canada, which has been the location of many murders and disappearances beginning in 1970.

from 1969 to 2006 (the RCMP also include women who have disappeared from Highways 97 and 5 in British Columbia). Ten of these eighteen victims are Indigenous women and girls. However, Indigenous groups argue that this number is misleading because it reflects only the disappearances and murders that have happened in the specific geographic areas around these highways and that the real number in Northern British Columbia exceeds forty. According to Human Rights Watch—an international non-governmental organization that conducts research and advocacy on human rights—British Columbia has the highest rate of unsolved murders of Indigenous women and girls in Canada.[2]

These numbers are startling. And many worry that these murders are going to remain unsolved because the victims are Indigenous. In fact, studies prove that Indigenous people across Canada are the victims of crime more often non-Indigenous groups. "In 2014, 28 percent of Indigenous people (aged 15+) reported being victimized in the previous twelve months, compared to 18 percent of non-Indigenous Canadians."[3]

Dubbed "The Highway of Tears," these murders and disappearances weigh heavy on the souls of those living in Northern B.C. If you flip through the pictures of the victims lost on the Highway, it comes into poignant focus that these were once vibrant young women with their lives ahead of them. To think how they were plucked away from their

families is a reality more harrowing than any horror movie. Some have never been found, while others were found too late, like twelve-year-old Monica Jack who was missing from 1978–1995, when her remains were discovered.

Filmmaker Taylor Sheridan was called to action by the murders on the Highway of Tears, as well as the unsolved murders of Indigenous women in the US. He based his 2017 film *Wind River* starring Jeremy Renner, Elizabeth Olsen, and Graham Greene, on the stark reality of violence toward women both on and off reservations. At the end of the film, before the credits roll, Sheridan reminds viewers that missing persons statistics are kept for every demographic in the US, except for Native American women. A chilling truth. There is also an award-winning documentary called *The Highway of Tears* (2015), narrated by actor Nathan Fillion, which focuses on the victims, as well as the generational poverty of the area.

It wasn't until 2002 when a hiker and, notably, a white woman, Nicole Hoar, went missing on the Highway that national media reported on the Highway of Tears.[4]

A bright light in the tragedy of the Highway of Tears is the formation of a task force in 2005 to finally find the perpetrators. At its height, about seventy investigators devoted their careers to providing justice for the victims' families. The task force, called E-Pana, has been successful in finding the murderer of sixteen-year-old Colleen MacMillen. Thanks to DNA, they were able to link American Bobby Fowler to her rape and murder, though he was dead by the time his involvement was uncovered. They believe Fowler could also be responsible for at least two more murders on the Highway of Tears.

E-Pana also saw to the conviction of Gary Handlen for Monica Jack's murder in 2019. He is also charged with the death of Kathryn-Mary Herbert. This work toward justice, in part thanks to the science of DNA, is revealing that the Highway of Tears was stalked by more than one serial killer. This led us to wonder if there is a geographical component to serial killing.

Statistics of US states have been collected for victims of serial murder from 1985–2010. New York wins the dubious honor with one hundred and thirty-seven victims. This is not too hard to believe, as New York City provides anonymity. Next comes California, the breeding ground for people like the Hillside Stranglers and Charles Manson. And in third place is Florida, where Ted Bundy finished his treachery. Most interestingly is the state of Washington, which for that time period had the highest per capita number of serial killer victims, 1.6 per 100,000.[5] After all that, it might seem that the United States has more serial killers than any other country. But Dr. Mike Aamodt, a forensic psychologist who studied five thousand serial killers around the world, begs to differ:

> In the United States, we have much more open records than other countries do. If the US had a higher murder rate than the rest of the world, I would be more likely to believe that we have more serial killers, too. But compared to other countries, in terms of the murder rate, we're right around the middle. First, law enforcement has to discover the murders and link them back to the same killer. That means you need competent law-enforcement agencies, which the US has. The second part of being able to track serial killers is once the killer is identified, it has to be announced by law enforcement and made available in prison records. The information must be available to the public.[6]

Okay, so maybe we have the most *recorded* serial killers, but not the actual most. England, South Africa, and Canada are at the top of the list under the US. It seems that American serial killers have just had more of

a chance to be depicted in the media. And as Dr. Aamodt explains, our law enforcement and record keeping only make it seem that we have an overwhelming amount. At least we hope that's true. Because although it might be rare to be killed by a serial murderer, they are real. They can climb into your window. They can stop, wearing a smile, and offer you a ride.

From 1980–1990, Alaska had the largest per capita of serial killer victims in any state (15.64 per million).[7]

The Highway of Tears spans hundreds of miles. It cuts through the land of many First Nations, carving its way to cities beyond. For too long, women unknowingly walked into a snare. Not just one man, but several, took it upon themselves to choose darkness over light. As the science of DNA blossoms, and investigative techniques improve, more light shines on the Highway of Tears.

CHAPTER TWENTY-FOUR
The Man from the Train (Unsolved)

One of the most compelling unsolved mysteries that I (Meg) have been drawn to over the years is the tragic murder of the Moore family in the sleepy town of Villisca, Iowa. The parents, their four children, and two visiting kids were all bludgeoned by an axe on the night of June 9, 1912. Someone moved from room to room, systematically snuffing out the lives of innocents. What haunts me is the senselessness. In the case of the famous Clutter family murders, documented in Truman Capote's *In Cold Blood* (1966), there is a clear motive of robbery. Even serial killers like Dahmer and Bundy have a "type" that they are motivated to kill repeatedly. With the Moore's murders, there is an overwhelming sense of randomness. They lived in an unassuming house in an unassuming

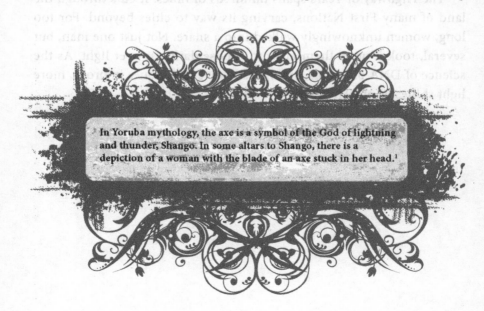

In Yoruba mythology, the axe is a symbol of the God of lightning and thunder, Shango. In some altars to Shango, there is a depiction of a woman with the blade of an axe stuck in her head.[1]

farm town. They were churchgoing, average American citizens with no untoward business practices or tantalizing secrets. This is the collective fear. That although we may live our lives with wholesome intentions, make few enemies, and mind our own business, the boogeyman can still come for us. He can creep into our homes where we feel the safest and take our lives with one, vicious strike of an axe.

The axe has long been a murder weapon in both brutal reality and the fake blood-soaked world of film. Stanley Kubrick famously altered the croquet mallet from Stephen King's novel *The Shining* (1977) into an axe for his 1980 film adaptation. In 1964's *Strait-Jacket*, Joan Crawford plays a woman who is accused of multiple axe murders, and let us not forget the horror comedy *So I Married an Axe Murderer* (1993), in which comedian Mike Myers plays a man realizing he may have been impetuous in falling in love with a mysterious woman (Nancy Travis) surrounded by a high body count. Perhaps Lizzie Borden is to blame for this fixation on the female and the axe, or maybe it is rooted in something more fundamental.

An article in the Chicago-based newspaper *The Day Book* from June 1912, depicting five of the victims from the Villisca axe murders.

In the ancient Minoan civilization on the island of Crete, the double-headed axe was a symbol of feminine strength. "It was the symbol of the Mother Goddess and signified the authority of women, matriarchy, and female divinities. A woman carrying a double-headed axe in the Minoan civilization indicated that she held a powerful position in society."[2] On the "Masonic Trowel," a website devoted to the Masonic brotherhood, axes are equated with duality.

The axe is one of the oldest tools of modern man. As with so many symbols, it has a dualistic association, in this case representing both destruction and creation. In prehistoric times, axes were made from stone, which sometimes created sparks. Many ancient cultures associated sparks with thunder, which in turn was known to have great powers. Native Americans, the Chinese, and even the Celts called axes "thunder stones." As such, axes became closely associated with power, both the power of destruction, and the power of creation.[3]

At the turn of the twentieth century, the axe was an object of industry. It was a tool of the lumberjack, the building pioneer, the industrious farmer. It created livelihoods, and in the case of numerous murders like the tragedy in Villisca, Iowa, it destroyed hundreds of lives.

Across the country, not long before the Moore family met their demise, the AxeMan of New Orleans stalked the French Quarter. The serial murderer, who was never caught, killed six, and injured yet another six with the blades of their own axes. He, like the Zodiac and BTK, enjoyed sending letters to the newspapers. What follows is a lengthy letter he hoped would strike fear into the heart of New Orleans:

Hottest Hell, March 13, 1919
Esteemed Mortal of New Orleans: The Axeman

They have never caught me and they never will. They have never seen me, for I am invisible, even as the ether that surrounds your earth. I am not a human being, but a spirit and a demon from the hottest hell. I am what you Orleanians and your foolish police call the Axeman.

When I see fit, I shall come and claim other victims. I alone know whom they shall be. I shall leave no clue except my bloody axe, besmeared with blood and brains of he whom I have sent below to keep me company.

If you wish you may tell the police to be careful not to rile me. Of course, I am a reasonable spirit. I take no offense at the way they have conducted their investigations in the past. In fact, they have been so utterly stupid as to not only amuse me, but His Satanic Majesty, Francis Josef, etc. But tell them to beware. Let them not try to discover what I am, for it were better that they were never born than to incur the wrath of the Axeman. I don't think there is any need of such a warning, for I feel sure the police will always dodge me, as they have in the past. They are wise and know how to keep away from all harm.

Undoubtedly, you Orleanians think of me as a most horrible murderer, which I am, but I could be much worse if I wanted to. If I wished, I could pay a visit to your city every night. At will I could slay thousands of your best citizens (and the worst), for I am in close relationship with the Angel of Death.

Now, to be exact, at 12:15 (earthly time) on next Tuesday night, I am going to pass over New Orleans. In my infinite mercy, I am going to make a little proposition to you people. Here it is: I am very fond of jazz music, and I swear by all the devils in the nether regions that every person shall be spared in whose home a jazz band is in full swing at the time I have just mentioned. If everyone has a jazz band going, well, then, so much the better for you people. One thing is certain and that is that some of your people who do not jazz it out on that specific Tuesday night (if there be any) will get the axe.

Well, as I am cold and crave the warmth of my native Tartarus, and it is about time I leave your earthly home, I will cease my discourse. Hoping that thou wilt publish this, that it may go well with thee, I have been, am and will be the worst spirit that ever existed either in fact or realm of fancy.
—The Axeman[4]

After this letter hit the press, the citizens of New Orleans crowded homes and bars, playing jazz throughout the night. It seemed to work, as no one was killed. That night.

What is the psychological impetus for taunting the police? Both the BTK Killer and the Zodiac sent letters, as did Jack the Ripper and the Son of Sam, while other serial predators provoked the authorities in other ways. The Hillside Stranglers, cousins Angelo Buono and Kenneth Bianchi, left the bodies of their victims on hillsides in plain sight near police stations. The Golden State Killer called his victims before and after their attacks, and also called the Sacramento Police to tease them about how he was the rapist they were after, and that he had already staked out his next victim.

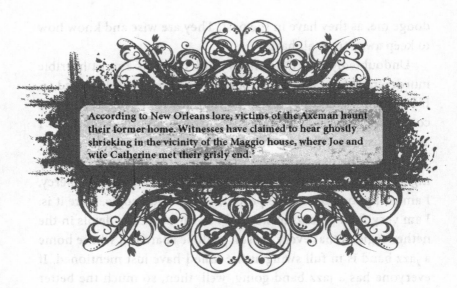

According to New Orleans lore, victims of the Axeman haunt their former home. Witnesses have claimed to hear ghostly shrieking in the vicinity of the Maggio house, where Joe and wife Catherine met their grisly end.[5]

One assumption could be that some serial killers share a strange desire to be caught. Criminologist James Alan Fox notes that these killers probably are not as self-destructive as they seem. "He (BTK) felt invincible. Unstoppable. And that's why many serial killers do communicate with the police. Not because they're looking for capture."[6]

Forensic psychologist J. Reid Meloy describes it as "both a desire to link with law enforcement and a desire to devalue it." On the other hand, the taunting could also simply be an expression of hubris if the criminal believes his system of communication is foolproof. Or it could be a need to ratchet up the risk because the contest with police has become as intriguing as the crime itself. Or even more simplistic, experts say, it could be a way of getting more attention. When they are caught, most serial killers usually have a stash of newspaper clippings about their crimes. The Unabomber followed his threat against LAX with the claim that it was just a prank to remind people that he was still around.[7]

Okay, so maybe these reasons aren't so nuanced after all. Like most compulsions of serial killers, this need to taunt seems to be one borne of a larger-than-life ego. Axe killing could be seen as an overt act of drama, too. Instead of killing with something more "quiet" like poison, or quicker, like a pistol shot to the head, axe murders are bloody, brutal, and painful.

Axe murders are aplenty, too, spanning the centuries and in countries all over the world. While the Villisca murders appear random in isolated research, patterns have emerged in the early twentieth century, specifically from 1898–1912, to indicate that the Moore family may have been the victims of a serial killer. Authors Bill James and Rachel McCarthy James posit this fascinating development in their book *The Man from the Train: The Solving of a Century-Old Serial Killer Mystery* (2017). Through tireless research, this father-daughter duo unravel a shocking series of axe murders that spanned across the United States and even in Canada. These murders share quite a few similar traits that lead many to now believe that a serial murderer was traveling the rails of America, on a mission to murder as many as he could.

One point of fact is that many axe murders during this era happened eerily close to a train track. This would be a convenient way for a murderer to hop off, commit a crime, and hop back on undetected. While numerous small American communities were grappling with the notion of a killer in their midst, it could be that the man who had taken an axe to an entire family (time and time again) was long gone. Because media traveled at a snail's pace mere decades ago, and also because a "serial killer" was a phrase not yet known or understood, it seems that this mysterious "man from the train" could be one of the most depraved and prolific serial killers to simply not be recognized as existing until he was long dead. One of my favorite aspects of the book is that at the end, Jones and McCarthy Jones actually reveal who they believe the "man from the train" is.

To learn more about *The Man from the Train*, as well as the axe as a murder weapon, we spoke with co-author Rachel McCarthy James, who's upcoming book *Whack Job*, explores the history and legend of the axe:

Kelly: "First, can you tell us what it was about this particular unsolved string of cases in *The Man from the Train* that compelled you to dive deep?"

Rachel McCarthy James: "We had an advantage in that we already knew there was a pattern found in a lot of crimes from 1911–1912. It was my dad's insight that there were crimes before those, and he found the first of them (in Hurley [VA] in 1909). He believed that the crimes were too practiced to be the work of a new killer, but he couldn't find anything in the years right before the 1909 event, so he hired me. At first, he just wanted me to look at 1907 and 1908, but I needed the money so I kept pushing the boundaries back. Within a couple of weeks I found the Lyerly murder, and then I kept finding more and more."

Meg: "One thing that really struck me about *The Man from the Train* was your commitment to facts. Some true crime books like to set a scene, speculate, even create dialogue to enhance the story. What led to you and your dad's decision to stay firmly in the corroborated facts?"
Rachel McCarthy James: "That's very kind of you to say! But I do think we wade a bit into speculation. We speculate on many things about his motivations and process, how he approached and chose families, how quickly he did his work—things no one could ever truly know. We also make guesses about how witnesses might have reacted. But we strived for clarity when departing from the events, as stated in our newspaper sources, and getting into our own reactions to the facts at hand, and tried our best to explain how we arrived at our conclusions.

The thing about the facts of this story is that, case to case, they tend to be pretty scant! There's a *ton* of information about Villisca, perhaps more on that one event than all the others combined. Some of the non-Villisca cases were well-covered by contemporary journalists, and the lynchings of Paul Reed and Will Cato have been consistently revisited and discussed in the century that followed. But some murders we have just a few hundred surviving words to go by—not even names sometimes. In the writing process, we referred to them as low-information events. So, when writing about those low-information events individually, we have to stick to just those facts because there are so few of them. There's just nothing more to go on, and almost any scene setting would cross the line too quickly. When these many less-described tragedies come together, there's a thudding insistence to them. The facts become hideously easy to anticipate, both for us as researchers and for you as the reader. We barely need to tell you that it was in the middle of the night with the back of the axe and the bedclothes piled on the body and the house locked afterward. It was true in the last chapter, and the chapter before that, and the chapter before that. We try to stick to the facts of all of these stories we found because, put together, they make their own argument."

Kelly: "Could you give us insight into your research process? What was it like pursuing knowledge from a century ago? And how did you feel when you found a link to or evidence of your theory? That had to be thrilling!"

Rachel McCarthy James: "I'd always enjoyed crime writing and true crime—I was obsessed with *In Cold Blood* [1965] as a teenager—and of course I'd had Harriet the Spy fantasies. But I could never have anticipated the moment I saw Paul Mueller's name, and the relief I felt when Dad, who is always quick to be contrary, saw what I saw. This wasn't just because I figured it out, but because I knew that meant this project would be a book, and that that book would also be my book.

And at the same time, no one wants to find more murders, especially not more lynchings, and they just kept coming. It's ghastly that it happened, and the excitement quickly wore off when we realized the mountain of work we had ahead of us. Especially since I wasn't just looking at the murders in this book—I was looking at hundreds and hundreds of different murders, trying to make sure we didn't miss anything (I tried to catalog every single murder of a family between 1890–1920). Of course, there's an emotional adjustment process, but it's a rough slog getting through it. And throughout the five-year-long writing process, there was the underlying anxiety that I might get scooped! Fortunately, that was unfounded.

My research process was basically to comb through newspaperarchive. com and find as much as I could on every crime we decided to include (and we erred on the side of inclusion). I revisited each crime in the book probably half a dozen times at minimum, trying to make sure I hadn't missed anything. We decided to only stick to newspapers with these crimes, no reporting trips. I think there's more to be found in local archives for sure, but with the volume of crimes we were working with it, would have been impossible."

Meg: "From Lizzie Borden to the Villisca murders, I, like many, am fascinated by axe murders. You explore even more about axes in your

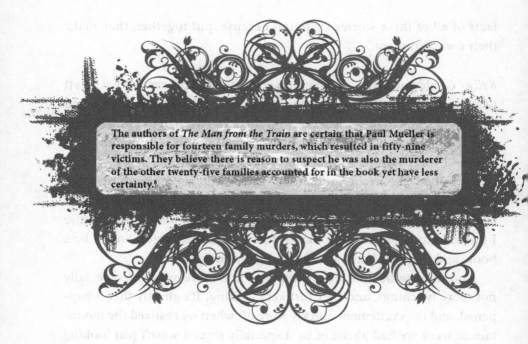

The authors of *The Man from the Train* are certain that Paul Mueller is responsible for fourteen family murders, which resulted in fifty-nine victims. They believe there is reason to suspect he was also the murderer of the other twenty-five families accounted for in the book yet have less certainty.[8]

upcoming book *Whack Job*. Why do you think we are so curious about this grim way to die?"

Rachel McCarthy James: "The axe is a primal and vital piece of technology, employed both as a tool and a weapon. People have certainly met violent ends with axes since before we were homo sapiens. Axes have always been around, available, and until very recently extremely commonplace, so it's always been a very convenient weapon, and thus it's played the instrument in many interpersonal murders—not to mention the heavy weight of state violence in executions and battle. An axe resonates deeply with our most base instincts. In today's world, where axes are less and less at hand and guns are everywhere, the idea of the axe is just a bit incongruous with our conception of murder. The idea of the "axe murderer" is just a little bit absurd! Which is why it was so effective in *The Shining* [1980], and why it became such a meme."

Kelly: "How would you compare the killer in *The Man from the Train* to modern day serial killers? Are there any other killers who have similar characteristics, whether in how they travel, or carry out their crimes, that you think would make a good modern comparison?"

Rachel McCarthy James: "Serial killers today are fundamentally different from serial killers in centuries past, because today we understand what serial killers are. In the 1910s, they didn't get that! The concept of someone targeting and murdering people for no good reason was completely alien to them."

Meg: "I think *The Man on the Train* could make for a fascinating film. Do you think it would translate well in this medium?"

Rachel McCarthy James: "Hopefully I'll have more to say about this soon (wink)!

Kelly: "Do you think there is a difference in the way the media handles "antique" murders versus modern murders? Is it easier for us to forget the humanity of the victims when they died so long ago?"

Rachel McCarthy James: "Hmmm. Sometimes I think the distance can actually help us see the tragedy more clearly. When people are reacting to a murder that's just happened, it's easy to project ourselves onto the victims or react to details taken out of context with authority we don't have. The media environment around murder can distort the facts as they emerge, and people start to pick and choose the details they promulgate or dismiss. Additional time grants us the perspective to see more of the story at once. Though, of course, many of the details are lost, which can be its own distortion. With a more recent murder, you have to be so careful because it's in part about people who are still living—the survivors of the victim, the witnesses, the accused—but it's easy to forget that, especially if it's in that not-especially-recent-but-not-forever-ago space, where curious onlookers are naturally drawn to the mystery and where the mystery can begin to take precedence over the living people involved. With older crimes, there's more of an opportunity for patience."

Meg: "Can you tell us more about *Whack Job*? What other projects can we look forward to?"

Rachel McCarthy James: "I'm so excited about *Whack Job*! It's a journey through our shared humanity—often at its worst, but also at its most creative, most vulnerable, most primal. Each chapter chronicles a different violent death—some from half a million years ago, some as recent

as 2019—and how that specific episode in brutality reflects the circumstances of the axe in the time and place where it happened. I'm immersed
in research right now and I'm learning so much—about violence, about
technology, about royalty, and evolution, and archaeology. I can't wait to
share it, hopefully in 2023. That's my primary project right now, but I've
got a ton of other long-term stuff in the works—and I might dive back
into the story of *The Man from the Train* at some point, too. (Wink!)"

It was an honor to speak with Rachel McCarthy James, and we are
waiting not-so-patiently to read *Whack Job*!

SECTION NINE
HOME INVASION

CHAPTER TWENTY-FIVE
The Strangers
(Charles Manson)

Home invasion is a terrifying prospect for anyone, regardless of the outcome. According to the Bureau of Justice Statistics, "an estimated 3.7 million household burglaries occurred each year on average from 2003 to 2007. In about 28 percent of these burglaries, a household member was present during the burglary. In 7 percent of all household burglaries, a household member experienced some form of violent victimization."[1] While only a small number of home invasions end in the unthinkable, they are still a real and palpable threat.

This is the premise of the 2008 film *The Strangers*. Director Bryan Bertino recalled reading *Helter Skelter* (1974) about the Charles Manson murders. "I was thinking about the Tate murders and realizing that these detailed descriptions had painted a story of what it was like in the house with the victims. But none of the victims knew about the Manson family or why it was happening to them. So, I got really fascinated with telling the victims' tale."[2] Charles Manson is the cult leader responsible for the murders of at least seven people including actress Sharon Tate.

Sharon Tate was an American actress and model who was murdered by members of Charles Manson's family in 1969.

How does one get involved in a cult and what is the psychology behind them? To understand more about this topic, we interviewed Dr. Janja Lalich, a researcher, author, and educator specializing in cults and extremist groups:

Kelly: "Tell us about your background and areas of expertise."

Dr. Lalich: "I was in a cult myself in the seventies and eighties. I had already had my college degree and all of that so when I got out after ten years, I tried to figure out what happened to my brain. Eventually I was better and after about ten years, I decided to go to grad school, and I got my PhD. I had already written a couple of books and was working with Margaret Singer, who was the preeminent cult expert at the time, and continued researching cults."

Meg: "Before you got into this cult did you know much about them?"

Dr. Lalich: "No. I was a French major for my bachelor's degree and then I got a Fulbright fellowship and studied another year in France. I came back; it was the seventies, I was a hippie. I didn't know much about cults until I got in one and of course I didn't think I was in a cult. Even though the newspaper in San Francisco would write articles about us saying we were a cult. We just did damage control when all of that happened. Because I had been in a political cult, and everything at the time was about religious cults, one of the first things I did when I got out was compare how we were a cult, given the criteria for religious cults. And yeah, we were in a cult. I started going to conferences and I was probably the first person that really talked about political cults."

Meg: "Charles Manson was a known cult leader. What can you tell us about his style of manipulation?"

Dr. Lalich: "I actually thank my lucky stars that I didn't get recruited by Charlie Manson because, when I was a hippie, I was traveling around in East Bay and San Francisco at that same time he was around. His main style of manipulation was sex and drugs. Most cults don't use drugs on their members. He would give these girls LSD and then have sex with them. That, to me, is the main way he manipulated people."

Kelly: "When researching the Heaven's Gate cult, I was astonished to realize that they often changed terms to reprogram the mind to disassociate from memories and regular thinking. Is this common in cults?"

Dr. Lalich: "Robert Lifton, in his research, uses the term 'loaded language.' I think all cults do that. They create their own terminology. They may even be the same words that we use but they have a different meaning. That has two purposes. It does what you were saying, it dissociates people from the world they had been living in before, and it's also a way to control them. It stops the critical thinking like when you hear that word you know exactly what it means, you know exactly what to think, you know exactly what to do. Over time it shuts down your critical thinking ability."

Kelly: "When did this science emerge about cults?"

Dr. Lalich: "Lifton studied the thought reform program in Communist China in 1961. Edgar Schein complimented Lifton's work with his book *Coercive Persuasion* (1961). Those works had been out there, but I don't think anybody applied it to cults until Margaret Singer, who had worked with Lifton and Schein. In the seventies, parents would come to her because their kids were being recruited by the Moonies or the Children of God and she remembered back to her work and thought these cults were like the Chinese thought reform programs. She was really the first one who brought that out."

Meg: "Is any one type of person more prone to becoming ensconced by a cult? Or are we all possible victims?"

Dr. Lalich: "In my opinion, everyone is susceptible. If there's any common denominator, it's idealism. Most people join a group because they think they're going to create a better world or a better life for themselves or maybe it's financial success or the path to salvation. It's an idealistic motivation on the part of most people. It's not because people are weak or stupid, which is what the myths are. And it's typically not the cult leader who does the recruiting. It's his or her members. A charismatic leader, if we want to call them that, just needs to get a couple of followers. Those followers keep recruiting and sending the message out. Most cult leaders are pretty lazy. They just like to sit back and take the glory."

Kelly: "In light of what happened in Washington D.C. on January 6, 2021, some people are calling Trump's following a cult. Do you agree with that?"

Dr. Lalich: "I think that a great majority, including the QAnon people, were in a kind of cultic relationship with Trump. And he certainly behaved like a cult leader; constantly riling them up, having those rallies, using buzz words, not allowing any criticism, getting rid of people in his administration who didn't fully support him. I think we definitely saw for the first time in our country's history a cult on a national scale."

Meg: "What else do you think is important for people to know about cults?"

Dr. Lalich: "One thing I think that's difficult for people to understand is this idea of charisma. We attribute charisma to people and that, from the get-go, is a power relationship. That's an imbalance of power. The one that you attribute the charisma to has power over you. It's this individual relationship and that's so much of what the hook is."

Dr. Lalich has written extensively about cults in her books, *Captive Hearts, Captive Minds* (1994), *Cults in our Midst* (1995), and *Bounded Choice* (2004). Thank you to Dr. Lalich for the fascinating interview!

"Charles Manson preyed on all aspects of the hippie zeitgeist to serve his purposes, from the Beatles to the commune where he would 'isolate [his followers] from their family and friends. Manson would give them hallucinogenic drugs and he would pretend he was also taking a drug, but he wouldn't. Then he would manipulate their trip and suggest things to them in a vulnerable state, guiding them through a journey where, at the end, they believed completely in him'."[3]

We see the end of the film first in *The Strangers*. We hear the recording of a 911 call after two boys discover murdered bodies in a house. What is the history of 911? The service wasn't invented until 1968 in the United States. Before that year, people who were experiencing an emergency needed to dial their local ten-digit number that corresponded to the agency they needed. Two years prior to the service being implemented, the National Academy of Sciences published a report that proved "accidental death and injury, particularly from motor vehicle crashes, had become an epidemic in the US. The report urged a series of steps to reduce these needless deaths and injuries, including exploring the 'feasibility of designating a single, nationwide telephone number to summon an ambulance.'"[4]

The film moves to a flashback and shows the couple before they met their demise. Kristen (Liv Tyler) and James (Scott Speedman) are on their way back to the cabin after a wedding reception. It's refreshing to see a horror movie cast people who are a little bit older, not the typical teenage types. It's also interesting to see that the couple is not in a perfect place in their relationship. Their strained rapport adds depth and dimension to their characters and makes the storytelling more complex. They have apparently not gotten engaged this evening, much to the chagrin of James. They have a reconciling of sorts a little after 4 a.m. but are interrupted by a knock at the door.

(Having lived in the country for most of my life, I (Kelly) can attest to how strange it would be to have someone show up and knock on your door in the middle of the night. We lived at the end of a long, dead-end driveway at least twenty minutes from the nearest town. When we saw headlights coming down our driveway at night, it alerted us that either someone was lost or someone we knew was coming over.)

The knock at the door belongs to a woman looking for someone named Tamara. Something I like about this scene is that we don't often find women to be scary. We don't assume that they will be an aggressor or a threat to us so having the first encounter with a "Stranger" in this film be a woman usurps our expectations.

James leaves Kristen alone in the cabin while he goes on a cigarette run. This is a more common horror movie trope; a woman alone equals potential danger. Kristen is alone in the cabin listening to music on a record player. Nothing could be more calming or soothing to the viewer. Yet we know, since this is a horror movie, that something bad is about to happen. She hears a bang on the door. Then another bang. After a gentle knock, the same woman asks through the door if Tamara is home. This is our first clue that something is not right here. One of the scariest non-jump scares in horror movie history occurs in this film. After a few more aggressive bangs on the door, a dead cell phone, and a fire mishap, a figure appears behind Kristen as she smokes her cigarette. The figure is masked and makes no noise. But the viewer knows that Kristen is in grave danger. By the time she turns around the figure is gone but we hear a noise and so does she. These moments build anticipation and fear in the audience.

During production of *The Strangers*, it was reported that Liv Tyler came down with tonsillitis due to the extensive screaming the role required her to do. Tyler would later recall it being the most difficult film she had ever worked on, both physically and emotionally.[5]

What is happening to our brain when we get scared? First, the amygdala scans for threats and signals the body to respond. Next, our brain stem may trigger a freeze response while our hippocampus turns on the fight-or-flight response. Our hypothalamus signals our adrenal glands to pump hormones and our prefrontal cortex interprets the event and compares it to past experiences. Finally, our thalamus receives input from the senses and "decides" to send information to either the sensory cortex (conscious fear) or the amygdala (defense mechanism).[6]

Some people enjoy the sensation, or emotion, of feeling fear while others do not. How you react depends on several factors including your past experiences with fear, your personality, the people around you, and the type of fear you're experiencing. For example, if you're attending a haunted house or watching a horror movie, you expect to be frightened. But being in real danger or peril will cause a different type of fear and you will react accordingly. Like the villains in the movie, some enjoy causing fear in others to witness their reaction or to assert power and control over them.

The couple can't escape with the vehicles on the premises. Could they if they had known how to hotwire a car? Maybe. Is hotwiring a vehicle as easy as movies make it seem? To the trained eye, it is possible to hotwire an older model of a car by removing the steering column cover and finding the wires that connect to the battery, ignition, and starter. Stripping the wires and twisting them together will start the car but this is not recommended unless you have experience (and, of course, you own the car)!

Kristin tries to use the Citizens band (CB) radio in the barn, but Dollface (Gemma Ward) smashes it before she's able to announce her location. What is the science behind CB radios? Citizens band radio originated in the United States in the 1940s and the FCC approved them for use as a hobby in the 1970s. They use forty channels and because of federal limitations, the range is no more than thirty miles. "Like other land mobile radio services, multiple radios in a local area share a single frequency channel, but only one can transmit at a time. The radio is normally in receive mode to receive transmissions of other radios on the channel; when users want to talk, they press a 'push to talk' button on their radio, which turns on their transmitter."[7]

One of the final scenes of the film is very reminiscent of *Halloween* (1978.) Kristen hides in a pantry with a slat door akin to Michael Myers (Nick Castle) trying to find Laurie Strode (Jamie Lee Curtis) in the famous slasher movie. She is indeed found and meets her fate alongside her lover, or perhaps now fiancé. She is now wearing the originally abandoned engagement ring as they declare their love for one another.

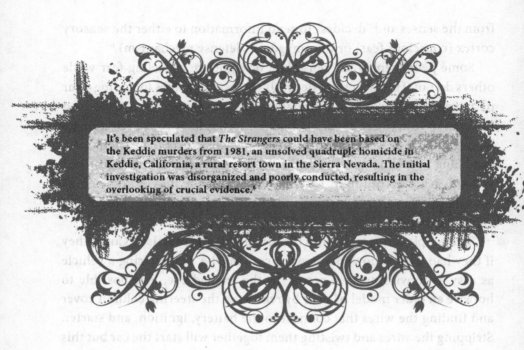

It's been speculated that *The Strangers* could have been based on the Keddie murders from 1981, an unsolved quadruple homicide in Keddie, California, a rural resort town in the Sierra Nevada. The initial investigation was disorganized and poorly conducted, resulting in the overlooking of crucial evidence.[6]

(Kristen doesn't die—she is alive when the boys discover the cabin later.) The killers' masks are removed but we, the audience, never see their faces. This makes the premise and conclusion all the more terrifying. Why did the killers do this? We may never know. Which is where the true horror lies.

CHAPTER TWENTY-SIX
Speck (Richard Speck)

On July 13, 1966, Richard Speck murdered eight student nurses in their home. This horrific and terrible crime shook the city of Chicago as well as the nation. To understand life in Chicago at that time we spoke to Robert Florence, my father-in-law, and former resident of Chicago during the time of the murders:

Kelly: "Describe what life was like for you, specifically, and Chicagoans in the summer of 1966."
Robert Florence: "In 1966 I was in college at the University of Illinois. I was off that summer from school, working in a factory in the plant where my dad was a research chemist. We were in the throes of the Vietnam War and that consumed the news. There was no thought at the time of a serial killer."

Meg: "Do you remember where you were and how you found out about the horrific murders that took place that summer?"
Robert Florence: "I do not recall exactly where I was when I heard the horrific news. But the local and national TV channels and the Chicago and national newspapers were replete with stories and details of the Speck murders. We were horrified about just how the murders occurred one by one with the nurses almost certainly knowing what their ultimate fates would be."

Kelly: "Did you, your family, or your friends follow the news about Richard Speck?"
Robert Florence: "My family and I, as well as the entire Chicagoland area and the nation, were fixated on the horrible news. It hearkened to the relatively recent news about the Boston Strangler (1962–1964), so I think that was fairly fresh in our minds. Nonetheless, the city was blindsided

that another serial killer in our town could happen so quickly in the aftermath of what happened in Boston. Most of us had long forgotten the physician who was a serial killer in the 1890s during the Columbian Exhibition. [H. H. Holmes] Ironically, that also occurred on the South Side of Chicago."

The Chicago Strangler is the nickname given to an American serial killer (or killers) suspected of raping and murdering at least fifty-five women and girls in Southwestern Chicago between 2001 and 2020. The killings were combined only in 2018, and until then they were considered to have been committed by different perpetrators. Nevertheless, representatives of the Chicago Police Department told the media in 2019 that there might actually be several serial killers operating in the city.[1]

Meg: "Did knowing about this terrible crime affect how you or your sisters lived your lives after that point?"

Robert Florence: "The news of Speck's murders did rattle us some since it reminded us of the evil that lurks among us. However, we lived in a quiet suburb on the Northwest edge of the city far from where the murders occurred. That area of the city was so different than the suburbs that it almost seemed nonexistent to us."

Kelly: "What stands out to you most about that summer?"
Robert Florence: "The pictures of Speck with his evil-looking, gaunt, pockmarked face were burned into my mind and the minds of others. To this day, I firmly believe that myself and others in the Chicago area could pick his picture out of a huge lineup. The other thing that I remember was the haunting tattoo that he sported on his arm: 'Born to Raise Hell,' and how prophetic those words were in coming to fruition. In the aftermath of the murders, the news was full of stories and interviews with the only surviving nurse, Corazon Amurao. We could all relate to the horrors that she must have experienced."

Amurao said in an interview fifty years later that "after that night, I'm always scared, you know."[2] She married and raised two children, one of which became a nurse herself.

A film version of the events of that night, *Speck*, was released in 2002. The actor playing Richard Speck (Doug Cole) is seen shooting up drugs in a bathroom stall very early on in the movie. It was reported that Speck, in reality, did heroin. How does this drug affect the body? Within moments after injection, heroin is turned into morphine by the brain. The brain's receptors receive it and release endorphins and dopamine, known as the "pleasure hormone." The rush of heroin has been described as feeling similar to a sexual orgasm, and can last up to two minutes. The rush is followed by a high that can last up to five hours. Heroin users then enter a state between sleep and waking that feels drowsy and warm. The brain adapts to these feelings, and when heroin isn't present, the brain and nervous system start to go into withdrawal.[3] Heroin rewrites brain chemistry, in a sense, and causes damage to our thinking and decision-making behaviors.

The film *Speck* focuses on the reactions of others rather than letting the audience be witness to acts of violence and rape. Much is seen in a fog or blur, seemingly the way Speck is viewing things or remembering them. The actresses perfectly capture the trauma that the real-life victims experienced. It lasted for a long duration of time and unfortunately, ended in death for all but one. How does this kind of experience affect survivors? Torture and the taking of hostages have existed for millennia. Recent studies show that both can produce post-traumatic stress disorder (PTSD) symptoms, many of which last a lifetime. Hostage-taking generates its own unique dynamics on the victims[4] wherein they can suffer from denial, impaired memory, shock, numbness, anxiety, guilt, depression, anger, and a sense of helplessness.

Richard Speck was identified through a police sketch that appeared in the evening newspaper after the murders. What is the science behind police sketches? They are a common practice during law enforcement investigations.

A forensic artist . . . usually interviews crime scene witnesses and victims about a perpetrator's appearance to create a composite sketch. The composite could be drawn completely by hand or computer generated. Sometimes, forensic artists use a combination of the two methods. But no matter how fine-tuned the police sketch methodology, the most crucial component of an accurate facial composite is an eyewitness's memory."[5] Our memories are not 100 percent accurate, though. We may remember things that fit our cultural view of the world or with unrecognized bias. The art and details behind police sketches are more complicated than one might think. When an eyewitness thinks they can't recall a memory, smell may trigger a response. If they remember the smell of the place or

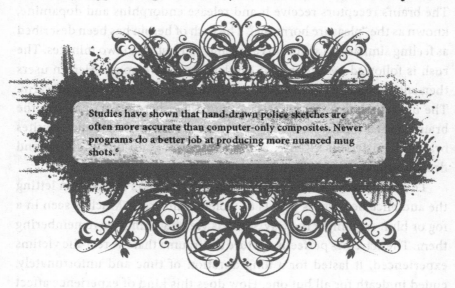

Studies have shown that hand-drawn police sketches are often more accurate than computer-only composites. Newer programs do a better job at producing more nuanced mug shots.[6]

the scent of the perpetrator, it can help bring back those visual memories as well. Also, not all police sketches come from memory. Some are filled in from surveillance videos or partial information. Although not always accurate, police sketches have led to the arrest of many notable criminals including Oklahoma City bomber Timothy McVeigh and Baton Rouge serial killer Derrick Todd Lee.

Speck viewed women as either perfect mothers or bad and promiscuous. This psychological condition is known as the Madonna-whore complex, as described on page 122. Men who hold these beliefs have an inability to view women as complex, independent people. This way of thinking "not only links to attitudes that restrict women's autonomy, but also impairs men's most intimate relationships with women."[7]

Speck was diagnosed with XYY syndrome which, in some cases, is believed to make males more prone to aggressive and violent tendencies. XYY syndrome is a "rare chromosomal disorder that affects males. It is caused by the presence of an extra Y chromosome . . . symptoms may include learning disabilities and behavioral problems such as impulsivity. Intelligence is usually in the normal range, although IQ is on average ten to fifteen points lower than siblings."[8] Speck is seen throughout the film changing his shirt after each murder and combing his hair. During his trial, it was speculated that he had obsessive compulsive disorder, and that this contributed to the events.

Many lines of dialogue in the voiceover narration in the film *Speck* were actually quotes from other serial killers, including Ted Bundy, David Berkowitz, Lucian Staniak, Ed Kemper, and Carl Panzram.[9]

Speck was also diagnosed with organic brain syndrome, resulting from the cerebral injuries suffered earlier in his life.[10] Most mental illnesses can be traced to biological or genetic factors but with organic brain syndrome, it can be traced to physical or emotional abuse. "The brain cells could be damaged due to a physical injury (a severe blow to the head, stroke, chemical and toxic exposures, organic brain disease, substance abuse, etc.) or due to psycho-social factors like severe deprivation, physical or mental abuse, and severe psychological trauma."[11] Symptoms of this condition include the inability to focus or concentrate, extreme rage or paranoia, and memory loss.

Speck claimed to have no memory of the murders. According to a 2014 study, amnesia for violent crimes is fairly common:

The results showed that amnesia for a violent offense was associated with crimes of passion and dissociative symptoms at the time, but not with impaired neuropsychological functioning. Long amnesic gaps were associated with a state of dissociation surrounding the offense and with previous blackouts (whether alcoholic or dissociative). Memory often recovered, either partially or completely, especially where there was a history of blackouts or a lengthy amnesic gap. Brief amnesic gaps were likely to persist, perhaps as a consequence of faulty encoding during a period of extreme emotional arousal (or red-out)."[12]

This is a controversial theory, and more research may be needed to truly understand the phenomena of blackouts.

Speck murdered eight women that night in Chicago, but their memories will live on. One victim's family has set up a scholarship in her memory. The Nina Jo Schmale Scholarship Fund has awarded twenty-five scholarships totaling $49,113 to date.

Richard Speck was convicted at trial and sentenced to death, but the sentence was later overturned due to issues with jury selection at his trial.[13]

CHAPTER TWENTY-SEVEN

American Horror Story: 1984 (The Night Stalker)

Between June 1984 and August 1985, a serial killer, serial rapist, and burglar terrorized the residents of the greater Los Angeles and San Francisco areas. The media dubbed him "The Night Stalker." His name was Richard Ramirez. A 2016 film version of his story stars Lou Diamond Phillips as the lead and he bears a striking resemblance to Ramirez. Phillips recalled, "I got to Los Angeles in 1986, literally a year after his reign of terror and after he'd been caught, so the specter of Ramirez was still there. That larger-than-life fear that gripped the community."[1]

Richard Ramirez was sentenced to death for his crimes but died in prison at the age of fifty-three.

How did Richard Ramirez get his start? When he was young, he bonded with a cousin over gory Vietnam photos. Are morbid fascinations normal or dangerous? Psychiatrist Carl Jung believed that our mental health depends on "our shadow, that part of our psyche that harbors our darkest energies, such as melancholia and murderousness. The more we repress the morbid, the more it foments neuroses or psychoses. To achieve wholeness, we must acknowledge our most demonic inclinations."[2] It's a natural reaction, beginning when we're children, to look toward the gruesome. It can build empathy or, in opposite extreme cases, tear down sensitivities.

A young Ramirez witnessed the murder of his cousin's wife. He didn't appear to have much, if any, empathy for the victim at his age.

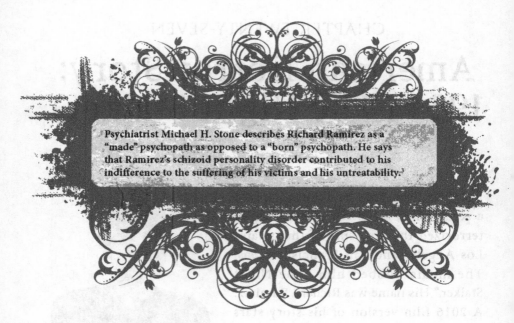

Psychiatrist Michael H. Stone describes Richard Ramirez as a "made" psychopath as opposed to a "born" psychopath. He says that Ramirez's schizoid personality disorder contributed to his indifference to the suffering of his victims and his untreatability.[3]

What could have contributed to this? Numerous studies have been conducted that show how children witnessing violent acts either at home, in the setting of war, or in their community, can have long-lasting, sometimes permanent, brain changes. According to the Child Witness to Violence Project, witnessing violence affects children's abilities to learn and affects their behavior. This may make it difficult to establish good peer relationships. Children who witness violence may be more aggressive, fight more often, and are at greater risk to become violent themselves.[4]

Footprint science helped identify Ramirez at crime scenes. How is this used by investigators? Forensic podiatry can determine a suspect's gait, footwear, and footprints. Unlike fingerprints, footprints themselves are not necessarily unique. "The use of foot-related evidence in criminal investigations dates back to 1862 when Jessie McLachlan's footprint placed her at the scene of a woman's murder for which McLachlan was subsequently convicted. In modern times, forensic podiatrists have assisted law enforcement in investigations since the 1970s."[5] To catalog the evidence of a footprint it should first be photographed then measured. If the footprint caused an impression, it should be cast. If possible, the

Maria Hernandez survived an attempted murder by Ramirez when a bullet fired at her ricocheted off the keys she held in her hands as she lifted them to protect herself.[5] But bullet ricochets can be lethal. A notable death caused by ricochet was the hostage Katrina Dawson during the Lindt Cafe siege in December 2014, killed by a ricochet from a police bullet when tactical officers stormed the building.[6]

surface that the footprint is on should be taken as evidence, like cutting out a piece of carpet or fabric. Shoes themselves can show wear patterns and these, too, can be detected from a print.

Richard Ramirez is a character in *American Horror Story: 1984* (2019) and is portrayed by Zach Villa. Although the events at the summer camp in the show are fictionalized, many of the scenes portraying his childhood are accurate. We spoke with Zach about his portrayal of Ramirez:

Meg: "How much did you know about Richard Ramirez before you were cast in this role?"
Zach Villa: "Basically nothing. When I got the audition, I didn't even know it was for *American Horror Story*. When I got the sides [sections of the script for the audition], I was like 'oh, there's clearly something off about this character. There's something sordid, and/or broken, and/

or lascivious about this person. I can't quite put my finger on it but I'm going to roll with that.' At the audition it became clear that they were directing me toward this person being very violent, broken . . . so I literally said, 'weirder?' and they said 'yes!' I then came to find out it was Richard Ramirez.

As far as the prep, I did seek out information. I read the book by Philip Carlo, *The Night Stalker* [1996], cover to cover and compiled all the information I could. What I mostly stuck with was what victims said about their experiences interacting with him and/or being attacked by him and his life story before he became this infamous serial killer. The progression of his crimes was interesting to me and obviously very important to see how his psyche and perversion developed. but what was more interesting was how he got there in the first place."

Kelly: "In my research for this chapter, I came across psychologists saying he was a 'made' serial killer, not a 'born' one. Did you also come across that?"

Zach Villa: "I did. That's a really interesting concept, nature versus nurture. I do agree that he was 'made.' I think that if you have a predisposition genetically to certain deficiencies or imbalances it can leave you open to developing a psychosis or perversion that you otherwise wouldn't. Certain things in his history . . . like, he didn't have a chance. It's obvious that these factors played into the development of the person he became and the behavior that he exhibited later."

Kelly: "Did you have any rituals at the end of the day to leave the character behind?"

Zach Villa: "For me, I think just getting off the set, getting out of the physical makeup and the physical language. The most realistic thing that codified for me was getting on my motorcycle at the end of the night and cruising home. That kind of visceral jump back into my environment wiped away any sort of memory of being in that place."

Meg: "What was your biggest takeaway from playing this role?"

Zach Villa: "The crew and the cast of *American Horror Story* in particular work so hard. It is a tireless and sometimes thankless job some days on set. We easily did fifteen-, eighteen-hour days a number of times. One in particular (**spoiler alert!**), basically right before my character dies the first time, that was one of the longest days I've experienced on set. My costar [John Caroll Lynch] and I were physically exhausted just trying to take care of our bodies and our brains and also make this really visceral, violent performance over and over and over again. There were a lot of techs involved but they work so hard and care so much at the end of the day about making really great entertainment.

Kelly: "Did you do anything differently to prepare for this role, given that the character was a serial killer?"
Zach Villa: "Anytime you make anything in film or TV or on stage and you're an actor, if it's based on a real person, you're making a version of that person. Sometimes you're going for an accurate portrayal of what they might have been like in real life but there are so many factors that adjust the tone of a performance. I was using real-life information and then doing a version of this role and of this person that fit into the world of the TV show."

It was absolutely fascinating to learn about the behind-the-scenes aspects of a television series. You can watch Zach Villa in season nine of *American Horror Story* or check out his band Sorry Kyle.

SECTION TEN
THE STRANGE

CHAPTER TWENTY-EIGHT
The Amityville Horror
(Ronald DeFeo Jr.)

When I (Kelly) was in the eighth grade, I read *The Amityville Horror* (1977) by Jay Anson over a couple of days. Growing up in a religious household, the book's themes of church and the devil terrified me. The book is based on the real-life murders that took place on November 13, 1974, in Amityville, New York. Ronald DeFeo Jr. shot and killed his parents and four siblings while they slept that night. At first, DeFeo claimed the murders must be linked to organized crime but throughout a night of questioning the picture unfolded and a confession was made. He said "once I started, I just couldn't stop. It went so fast."[1] What should have been an open-and-shut

case took a twist, though. DeFeo soon claimed that he had heard voices in the house that directed him to kill. This supernatural element that plagues the home is the focus of the book and the 1979 film.

The creepy children's voices singing "la las" over the opening credits while the "eyes" of the house peer at us through the changing light of dusk brings a literal chill to my bones whenever I rewatch this movie. The

Serial killer Ronald DeFeo Jr. murdered his family.[2]

score was nominated for an Academy Award composed by Lalo Schifrin. The film transitions into a stark contrast with dark, thunderous rain and booming lightning, seemingly in perfect timing to the gunshots we see and hear through the windows of the house.

Shopping for a new home can be an absolutely thrilling but stressful experience. Imagine touring such a luxurious and beautiful home and not knowing the history that it holds. Does it matter? For some people, who may be superstitious or believe in the supernatural, the past of the home absolutely matters. For others, it may not matter so much. It did end up mattering to the Lutz family. They reportedly started experiencing odd occurrences not long after they moved in. Drafts of unexplained wind are featured throughout the film. The family, as well as visitors, feel this strange occurrence. What could cause this in real life? Home inspection experts would point to air leakage issues in the home. These could include drafts coming through doors, windows, power outlets, or even cable TV and phone lines. An energy audit can be conducted including a blower door test, which depressurizes a home, to reveal the location of any leaks.[3] Supernatural experts would point to other possible explanations for these phenomena. Strong "paranormal winds" have been reported to take place preceding, or during seances, present during hauntings, and felt during visions of apparitions. Emanuel Swedenborg, a seer in the 1700s, wrote in his spiritual diary, "a spirit is compared to the wind; hence it is that spirits have come to me both now, and very frequently before, with wind, which I felt in the face; yea, it also moved the flame of the candle, and likewise papers; the wind was cold, and indeed most frequently when I raised my right arm, which I wondered at, the cause of which I do not yet know."[4] Paranormal researchers over the centuries have reported feeling these types of winds when investigating homes or structures where hauntings have been reported. A laser grid scope can detect any breaks in pattern in a space and are used by modern ghost hunters along with other electronic gadgets.

Related to this are perceived temperature differences in spaces. The family in *The Amityville Horror* experiences feeling cold in certain places or at particular times. According to ghost hunters, a cold spot is an area of localized coldness or a sudden decrease in ambient temperature. Many ghost hunters use digital thermometers or heat sensing devices to measure such temperature changes.[5] Believers claim that cold spots are an indicator of paranormal or spirit activity in the area; however, there are many natural explanations for rapid temperature variations within

"A full 49 percent of prospective home buyers said they would not consider a haunted house under any circumstances, regardless of price cuts or added perks. But 18 percent of respondents said the perception of a house being haunted wouldn't factor into their decision-making, nor would they need a concession to buy. The rest said they'd need some type of perk: a lower price, a larger kitchen, or a better neighborhood."[6]

structures, and there is no scientifically confirmed evidence that spirit entities exist or can affect air temperatures.

Kathy Lutz (Margot Kidder) asks a local priest to come over and bless their new home. When the priest arrives, he experiences seeing the gathering of the flies in the window, the door slamming shut by itself, and a growing sense of unease. Flies gather on the priest's face at an alarming rate until he hears a disembodied voice tell him to "get out." As unbelievable as this scene may be, some bugs are attracted to certain people. An individual's genetics affect our skin temperature, humidity profile, and metabolic rate.

Metabolic rate influences local carbon dioxide levels, which along with ammonia and lactic acid and other aliphatic carboxylic acids influence landing rates of biting insects like mosquitoes. Each human thus has a largely individual VOC profile, a product of their unique genetics and unique skin microbial profile. In turn, biting insects have each their own specific odorant receptors. A combination of these two parameters likely makes some humans more attractive to each such biting insect compared to others.[7]

Our diets may also contribute to an insect's preference for biting someone or not. Further research in this area is needed (but you won't find me signing up for that study!).

The priest also experiences the wounds of Jesus on his hands. How common has this been throughout history? This phenomenon is known as stigmata and was mentioned in the Bible in Galatians 6:17. This is considered a "Holy Wound," one that Jesus was inflicted with during his crucifixion. The first recorded case took place in 1224 to St. Francis of Assisi while he was on a forty day fast. He saw an angel and experienced wounds in his hands, feet, and side. In the centuries since, doctors have purported that St. Francis may have suffered from quartan malaria or even leprosy, which would explain hemorrhages of the skin and intense pain.[8]

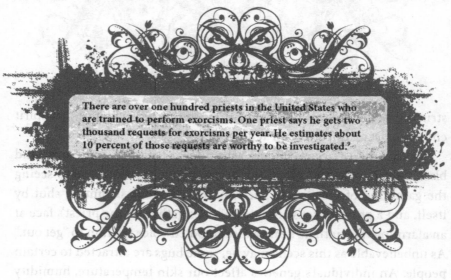

There are over one hundred priests in the United States who are trained to perform exorcisms. One priest says he gets two thousand requests for exorcisms per year. He estimates about 10 percent of those requests are worthy to be investigated.[9]

A common theme in horror movies is a family's dog sensing danger or paranormal activity. Is this possible? Scientifically, it could be explained by dogs' better sense of sight, sound, and smell. A dog's sense of smell is ten thousand to one hundred thousand times better than a human's.[10] Dogs have been known to detect cancer in humans and even detect pregnancy. Service dogs can read body language to understand mood and predict oncoming seizures or other medical conditions. Others suggest that perhaps we all have a sixth sense when it comes to danger but humans don't tap into theirs. Pet psychologist, Marti Miller, said "humans judge or deny what they're feeling. Dogs don't judge what is going on in the environment. While our own minds start to analyze what is happening, dogs don't do that. They feel the barometric pressure change, and may react by shaking, panting, salivating, and feeling anxious, or they may not react at all."[11]

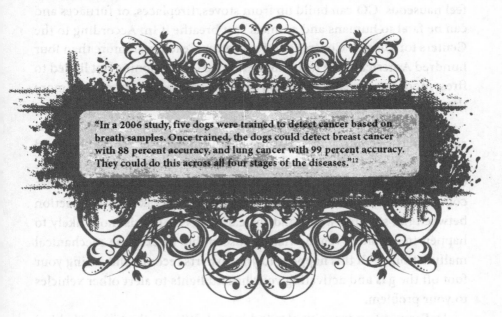

"In a 2006 study, five dogs were trained to detect cancer based on breath samples. Once trained, the dogs could detect breast cancer with 88 percent accuracy, and lung cancer with 99 percent accuracy. They could do this across all four stages of the diseases."[12]

In the opening police investigation scene in *The Amityville Horror*, experts placed the time of the murder between 3 a.m. and 3:15 a.m. On the first night that we see in the house, the husband wakes up at exactly 3:15 in the morning. What is the science behind waking at this time? According to sleep experts, waking up at 3 a.m. is common and has a reason rooted in history. With the advent of artificial light, humans were able to stay up later into the night and disrupt their natural circadian rhythm sleep cycle. Prior to the Industrial Revolution, sleep was segmented into two parts in Europe and North America: first sleep and second sleep. Later, sleep has shifted to be in one segment in order to maximize work time. Sleep experts also cite the fact that by 3 a.m., most people are exiting the deepest cycle of sleep, REM, and naturally wake up more often for the remainder of the night.[13] Dr. Wayne Dyer, a self-help author, said that if you wake up at 3 a.m., "don't go back to sleep. These hours before dawn are when you are close to Source, and a great time of inspiration and creativity. Put your feet on the floor, get out of bed, feel the morning breeze, and listen to your inner thoughts."[14]

Both the priest and the nun retched after leaving the house. Carbon monoxide (CO) is the most common and likely culprit in making people feel nauseous. CO can build up from stoves, fireplaces, or furnaces and can be fatal to humans and animals that breathe it in. According to the Centers for Disease Control and Prevention, "each year, more than four hundred Americans die from unintentional CO poisoning not linked to fires, more than twenty thousand visit the emergency room, and more than four thousand are hospitalized."[15]

When the priest tries to return to the Amityville house, the car he is riding in loses control. the driver is unable to turn the wheel, the hood flips up, and they crash. Although total steering failure is rare in vehicles, it is possible. "Total steering failure whereby the steering becomes completely unresponsive can only be caused by a break or disconnection between the steering wheel and the car's wheels. This is more likely to happen in an older vehicle that is susceptible to extreme mechanical malfunctions."[16] If this happens to you, experts recommend easing your foot off the gas and activating your hazard lights to alert other vehicles to your problem.

DeFeo used an insanity plea in his trial. What is the history behind this? The M'Naughten standard "is the classic example of the insanity defense. It originated in Britain where, in 1843, M'Naughten murdered the secretary of the Prime Minister (in an attempt to kill the Prime Minister) because he believed that there was a conspiracy against him involving the government. The high court found him insane, and he was hospitalized. The court described what is now known as the M'Naughten standard, and in simplified form it says that at the time of the act, the person had a mental disease or defect that interfered with his ability to understand the nature and quality of the act he was performing or if he knew so, he did not know it was wrong."[17]

The father in the film starts showing signs of paranoia. "Paranoia is the feeling that you're being threatened in some way, such as people watching you or acting against you, even though there's no proof that it's true. It happens to a lot of people at some point. Even when you know that your concerns aren't based in reality, they can be troubling if they happen too often."[18] Symptoms may include being defensive, hostile, aggressive, easily offended, and not being able to trust others.

Over half of Americans believe that houses can be haunted. To understand more about this phenomenon, we interviewed radio host on Darkness Radio, Dave Schrader, who has appeared on *Ghost Adventures* (2008–) and other paranormal television shows:

Kelly: "Tell us about your background and training for going into haunted properties."
Dave Schrader: "There was no formal training for investigating claims of the paranormal, at least not for me. I am a lifetime experiencer and have had many different connections with the paranormal/supernatural. I decided to actively pursue my interests and fascination with history and the hauntings associated with them by just going out, putting myself *in* the locations, familiarizing myself with the history, and letting the haunting do the rest. It was much more about being in the moment for me than it was to try to break it down and analyze it. I just wanted to see it for myself."

Meg: "The family in *The Amityville Horror* come across places in their home that feel cold. How do experts explain cold spots in homes?"
Dave Schrader: "Cold spots are associated with most haunted locations. The theory is that the energy it takes for a spirit to manifest actually creates the cold spot."

Kelly: "What are your experiences with house blessings, cleanings, or exorcisms?"
Dave Schrader: "As an investigator, we try to honor the belief system of the homeowner or experiencer in order to help clear unwanted energy from a location. If you practice a religion other than Christianity, I will try to contact a holy person from that field of belief you ascribe to. I

get the basics from them and try to empower the experiencers to do the clearing and take back the space. I find sometimes that setting that intention and following through is enough. I personally will follow up with a prayer or blessing I am familiar with to try to seal the home after the victims of the haunting have done their part. I believe personally that most spirits do not want to scare, and once they realize they are unwanted, they leave or go about their afterlife in other ways without drawing as much attention. The most important element in dealing with clearings is the active participation of the people being affected. Arming them with knowledge and confidence is paramount in the first steps. We all need to realize we are not weak, and we too wield power that can establish clear boundaries."

Meg: "The priest in the film witnesses flies gathering in an upstairs bedroom as he attempts to bless the home and begins to feel physically ill. Have you encountered any strange phenomena when entering a home or property?"

Dave Schrader: "Yes, and I was actually knocked to the ground while investigating the Whaley House in San Diego. It was a powerful energy and moment for me. Instead of leaping to the conclusion that it was a demon or dark force, I realized that the spirit we were most likely interacting with was just angry, hurt, and showing a sign of power because of how his life and legacy were taken from him. I showed respect and empathy and had no more bad encounters there.

With that said, I have entered buildings that left me feeling sick to my stomach or brought on massive headaches. If the spirits are manipulating energy or the location is on a ley line or perhaps has been the focus of many terrified people going in and out over the years, they may have imprinted on the location and you are experiencing a by-product of that energy and fear. If you are more tuned in than others, that energy may very well affect you. It is always important to examine a location in daylight and always with a partner to get a good look and understanding of any potential dangers, like loose wiring that can cause shocks or emit high levels of electric and magnetic field (EMF) energy that can disrupt you and make you feel sick, edgy, or even cause you to hallucinate. It is also good to get a thorough look at it in daylight to see if there are broken windows, holes in the floors, walls, or ceilings that could pose potentially dangerous falls or unexpected breezes.

Always be well hydrated because dehydration can cause dizziness, headaches, stomachaches, brain fog, energy drain, all symptoms often associated with paranormal activity. Avoid stimulants like high doses of caffeine and nicotine as those are known to cause hallucinatory issues as well. Be honest with yourself. If you are in a bad mood, emotional, etc. before entering a location, that can also give you false experiences you may attribute to the paranormal."

Kelly: "Is there anything else you think readers would be interested to hear about regarding these topics?"
Dave Schrader: "The afterlife is a question everyone wants immediate answers to, but there is no immediate answer. Be respectful to the living and the dead, they were people too."

Ronald DeFeo Jr. may or may not have experienced supernatural beings during his time at the Amityville house, but perhaps if he had treated humans with the same reverence he had treated the voices he heard, his family's story may have had a happier ending.

CHAPTER TWENTY-NINE
The Boston Strangler (Albert DeSalvo)

When *The Boston Strangler* premiered in 1968, movie critic Roger Ebert was not a fan: "The problem here is that real events are being offered as entertainment. A strangler murdered thirteen women and now we are asked to take our dates to the Saturday night flick to see why. Gerold Frank's original book was written with the most honorable intentions, I believe, but the movie is something else: A deliberate exploitation of the tragedy of Albert DeSalvo and his victims."[1] Tony Curtis stars as DeSalvo, the serial killer who murdered thirteen women in Boston in the 1960s.

Parapsychology and extrasensory perception were used in police investigations for this case. It was a controversial decision at the time, but police have been known to use psychics over the past several decades. What is the history behind this process? The College of Policing released a document warning against using psychics if it will distract from the investigation itself. "High-profile missing person investigations nearly always attract the interest of psychics and others, such as witches and clairvoyants, stating that they possess extrasensory perception. The person's methods should be asked for, including the circumstances in which they received the information and any accredited successes."[2] Numerous scientific studies have been conducted to test the accuracy of psychic abilities in investigations, but none have shown definitive proof of the practice being legitimate. Psychics helping investigate strange phenomena have been portrayed in horror movies including *Insidious* (2011) and *The Frighteners* (1996).

There were several key facts that were similar in all the cases of the Boston Strangler. First, his victims were all women. Second, there were no signs of break-in so the victims themselves must have let the

"The Cold War created paranoia that the Russians might have a monopoly on individuals with psychic abilities working as government intelligence agents, so the US felt the need to pursue the same. A resulting undercover army unit was created in the early 1970s. The program was finally dismantled and declassified in 1995."[3]

perpetrator in. Third, all the crimes apparently took place during the daytime. Finally, there were no signs of struggle and no physical evidence found for any of the murders. Is this rare? DeSalvo was convicted of the crimes based on his confession, and later, DNA evidence, but usually there needs to be more evidence. The "corpus delecti" rule, which translates to "body of the crime," refers to "the requirement that there be some kind of evidence, apart from the defendant's statements, that establishes that someone committed a crime. In some states, the prosecution can't even present evidence of the defendant's confession (for example, by playing a recording of it) without this kind of corroboration."[4]

DeSalvo was considered a "power assurance" motivated rapist. They are prone to taking credit for crimes and don't consider rape equal to sex. The signs of this type of rapist include "attacking to reassure himself of his masculinity by exercising power over women, lacking confidence in his ability to interact socially and sexually with women, utilizing 'surprise' approaches and may claim to have a weapon, and often described as 'polite.'"[5] These types of rapists tend to be insecure, are socially awkward, and often take personal items from victims. It's important to note that rape, in general, is not about sex but instead about power and control. Statistics show that 45 percent of rapes are perpetrated by acquaintances of the victim.

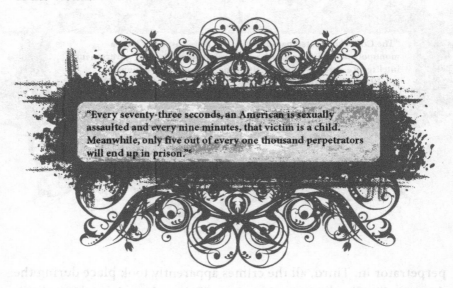

"Every seventy-three seconds, an American is sexually assaulted and every nine minutes, that victim is a child. Meanwhile, only five out of every one thousand perpetrators will end up in prison."[6]

In July 2013, DNA was matched between seminal fluid found at the rape and murder of Mary Sullivan and DNA obtained from DeSalvo's nephew, linking DeSalvo to the murder of Sullivan and excluding 99.9 percent of the remaining population.[7] DeSalvo's body was exhumed, and it was confirmed to be a DNA match. To understand more about how DNA is used in investigations, we spoke to Lisa Schliebe, a forensic DNA criminalist:

Albert DeSalvo was never convicted of his strangulation crimes, but it was proven after his death that they were committed by him.

Kelly: "Movies and fictional TV crime shows routinely use DNA evidence to prove the guilt or innocence of the accused. How accurate is DNA science and the techniques portrayed fictionally compared to real life?"

Lisa Schliebe: "In most cases, whether portrayed in novels or on TV, the science is about 50 percent accurate. Many of the techniques for collection and testing are well known and are portrayed well. However, the time frame that it takes to get a result is usually skewed for drama purposes. The involvement of the scientist in the case, usually alongside the detective, is greatly exaggerated as compared to real life. The biggest part that always seems to be missing is what happens in court. Having the scientist on the stand and explaining to a jury what tests were done and what the results mean are a large portion of our job. That is a part you rarely read about or see."

Meg: "Can you explain to us, and to our readers, how DNA science came about? And how do you see the science advancing in the next decade or so?"

Lisa Schliebe: "The earliest techniques were based on serology methods such as blood type, and we needed large drops of blood, roughly the size of a quarter. As we learned more about the structure and how it replicates, the methods became what they predominantly are now which is copying specific repeated segments that allow us to develop a DNA profile. We are now able to do this with a drop so small, it can fit on the head of a pushpin. The science is already advancing beyond what we thought possible as companies are developing ways to see the full DNA sequence of a person rather than small repeated segments. This change brings into play the use of ancestry information and physical traits like hair and eye color as ways to identify persons involved in a crime. I think that aspect may have the greatest benefit to cold cases where we've exhausted every other option. I see the current companies developing more sensitive kits allowing us to test more and get better results. We've already come a long way, but there are always advances to be made."

Kelly: "How long does it really take to process DNA evidence and what does it entail?"

Lisa Schliebe: "Let me start by explaining what is normal as it pertains to processing casework and then I'll tell you what is possible, but only on the rarest of occasions. The actual processing of DNA in the lab takes about two to three days on average. Why it takes so long to get report results is because of all the before and after activities. The evidence has to be screened for possible biological fluid and samples taken before the DNA process starts. After the lab work is the interpretation, which can take the longest time. A lot depends on the quality of the evidence and if we have access to reference samples to compare to. After this, the report is written and then the case as a whole goes through two layers of review before the report is made available to the investigating officer. For my agency, we have a standard ninety-day turnaround time. The clock starts once the request is received and reviewed by our unit and we've verified that all evidence is booked and ready for analysis. Now, as far as what is possible, when we have situations of increased danger to the public, such as an active serial case, we have the ability to turn around a request in a few days. As I mentioned, that is on the rarest of occasions, but has been done."

Meg: "For those who are interested in pursuing a career like yours, what advice would you give them?"
Lisa Schliebe: "You must have an undergraduate degree in a physical science (biology, chemistry, physics, etc.). After that, it is highly

George Nassar, the inmate DeSalvo reportedly confessed to, is among the suspects in DeSalvo's case. He is currently serving a life sentence for the 1967 shooting death of an Andover, Massachusetts, gas station owner named Irvin Hilton.[8]

recommended that an internship is completed and that a graduate degree is achieved. While these are not required, the field has become extremely competitive and most of the applicants have advanced degrees and/or work experience in science labs. Anything you can do to make yourself stand out is the best advantage to have."

Great advice for those interested in pursuing this important area of science and crime investigation!

Because there was no physical evidence at the time and he did not match witness descriptions, DeSalvo was never tried in any of the "Boston Strangler" murders. He was, however, sent to prison for life for the rapes and sexual assaults from his past. He was sent to prison in 1967 to serve his sentence, but six years later he was stabbed to death in his cell.

CHAPTER THIRTY
Eaten Alive (Joe Ball)

I (Kelly) heard a story, as a child, of a family friend who was chased by an alligator in Florida. She remembered a tip: run in a zigzag pattern. An alligator won't be able to turn as quickly as you can and you'll be able to escape! She did, indeed, usurp the alligator but fell and broke her arm in the conquest. I remembered this piece of advice throughout my life and

"Gators have a bite strength of 2,125 pounds per square inch—enough to bite through steel. The saltwater crocodile can slam its jaws shut with a force of 3,700 PSI."[1]

have, thankfully, never needed to use it. The horror movie *Eaten Alive* (1976) doesn't feature anyone being chased by an alligator but, instead, focuses on a real-life serial killer who may have used alligators to hide evidence of his crimes. Joe Ball is the suspected serial killer who may have murdered as many as twenty women in 1930s Texas.

A young Robert Englund, who horror fans will recognize from *A Nightmare on Elm Street* (1984), plays Buck in the movie. Englund, a classically trained actor, was excited to work with multiple legendary actors on the film. Walking onto the set the first day he recalled thinking, "God this is gonna be a great movie. I'm so glad I said 'yes.'"[2] He was also excited to work with master director Tobe Hooper. Like *The Texas Chainsaw Massacre* (1974), Hooper explores the terror that lurks in seemingly wholesome settings. The film relishes showing abandoned or derelict scenery. These settings have come to be synonymous with strange, creepy, loner types. Water-stained walls create an atmosphere of rotting and neglect while country music creates the juxtaposition of happiness and innocence. The soundtrack for *Eaten Alive* is similar to *The Texas Chainsaw Massacre*. Both were composed by Tobe Hooper and

Wayne Bell. They use real elements and sounds found in the environment to create an unsettling atmosphere. Happy, upbeat music plays within the world via the radio or record player while ominous, dread-building tracks underscore and help create tone.

Joe Ball was a bootlegger during Prohibition. "Bootlegging helped lead to the establishment of American organized crime, which persisted long after the repeal of Prohibition."[3]

In one scene, a man behind a desk turns from jolly to intimidating when learning that the young woman he's speaking to is a sex worker. The man is Judd (Neville Brand), who is a fictional version of serial killer Joe Ball. Ball has gone down in history as a folklore figure in Texas. The strange stories behind Ball's killings include bodies being disposed of in barrels and others fed to alligators. The latter is what the film version of his life focuses on. Part of the legend of Ball includes the story that he built a pond with alligators because he misunderstood the term "corpus delicti," a legal term meaning a murder conviction without a body would be impossible. Whether this is the real reason Ball kept alligators is speculation. But the legal term is much clearer. "The 'Corpus Delicti Rule' does not mean a dead body must be produced to prove a murder took place, but it does mean that the prosecution must produce some independent evidence of a crime before being able to introduce a defendant's confession or admission to a crime."[4] Ball was never formally charged so this rule is a moot point.

Have people ever used alligators or other animals to dispose of bodies? Disturbingly, the answer is yes. In 1902, a hog farmer named Joseph Briggen was sentenced to life in prison for murdering at least a dozen people and feeding them to his pigs. He claimed that human remains made his pigs grow larger and taste better. In 2007, a jury convicted Canadian serial killer Robert Pickton of killing and feeding forty-nine sex workers to his hogs.[5]

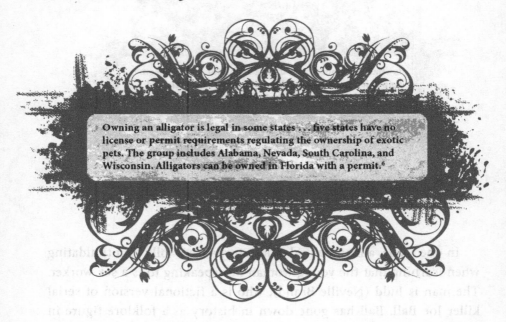

Owning an alligator is legal in some states . . . five states have no license or permit requirements regulating the ownership of exotic pets. The group includes Alabama, Nevada, South Carolina, and Wisconsin. Alligators can be owned in Florida with a permit.[6]

Whether Joe Ball actually fed his victims to his alligators, or crocodiles in the film, it got us wondering . . . what do crocodiles actually eat? And how long does it take for them to consume something? Crocodiles can, and sometimes will, eat anything they encounter. "A croc's stomach is the most acidic of all vertebrates, allowing it to digest bones, horns, hooves, or shells. Nothing gets left behind in a crocodile's dinner."[7] The dog in this film dies by getting eaten by a crocodile but does this happen often? In 2019, it was reported that twelve dogs were attacked in one county in Florida that year.[8] There are several hundred crocodile attacks per year in Africa alone and it is said to be far more likely to get attacked by a crocodile than a shark.[9]

Hattie (Carolyn Jones) calls the animal an alligator while Judd refers to it as a crocodile. What is the difference between the two? The snout shape and jawline are the easiest way to spot the difference between the two animals; alligators have a u-shaped snout while crocodiles have v-shaped ones. Alligators are opportunistic feeders, they typically won't chase people down unless provoked, while crocodiles tend to be more aggressive and therefore dangerous. (That must mean our family friend was provoking the alligator that chased her! The plot thickens.)

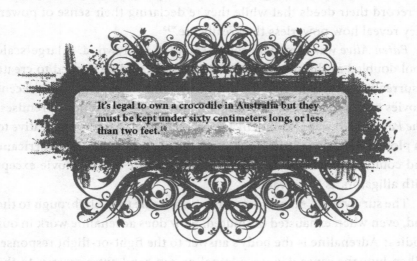

It's legal to own a crocodile in Australia but they must be kept under sixty centimeters long, or less than two feet.[10]

The father and sister of the first victim help search for her. How often do families get involved in missing persons cases? It is quite common for families to be directly involved in these cases, according to the National Center for Missing and Exploited Children:

Nearly 800,000 children are reported missing each year . . . 203,000 children are kidnapped each year by family members. Another 58,200 are abducted by non-family members. Despite these huge numbers, very few children are victims of the kinds of crimes that so often lead local and national news reports. According to NCMEC, just 115 children are the victims of what most people think of as "stereotypical" kidnapping . . . of these 115 incidents, 57 percent ended with the return of the child. The other 43 percent had a less happy outcome.[11]

The first forty-eight hours of a person missing due to possible crime are the most crucial and families are encouraged to work closely with law enforcement officials to help locate their loved one.

Judd keeps notes of all his victims in the film. Is this common for serial killers? Throughout history, diaries or notes written by serial killers have been found. Forensic psychology professor Dr. Katherine Ramsland said "those killers who want their secrets recorded take a risk. Some simply don't believe they'll be caught, but others might have such a compulsion to record their deeds that while they're declaring their sense of power, they reveal how powerless they really are."[12]

Eaten Alive was filmed on sound stages in Hollywood. A large-scale pool doubled as the swamp and Tobe Hooper said he wanted to create a surrealistic, twilight world[13] with the sound stage. Two other recent movies explored killer crocodiles with interesting and unique premises. *The Pool* (2018) was also shot in a large-scale pool that was imperative to its plot while *Crawl* (2019) took place during a category five hurricane and could be described as a home invasion-style horror movie except with alligators as the invaders.

The survivors in many horror movies are able to push through to the end, even when exhausted or injured. How does adrenaline work in our bodies? Adrenaline is the body's answer to the fight-or-flight response. When humans sense danger, adrenaline can send extra oxygen to the lungs, decrease the body's sense of pain, increase physical strength, and sharpen focus.[14] People have been known to accomplish extraordinary feats with a surge of adrenaline. In 2006, a man was able to lift a three-thousand-pound car off of someone[15] while that same year a woman fought off a polar bear until it was shot by a local hunter.[16]

Judd bites it in the end of *Eaten Alive* by getting bit by one of his alligators, but what happened to the real killer? Joe Ball died by suicide in 1938 after being approached by two sheriff's deputies. He is confirmed to have killed two women, Minnie Gotthardt and Hazel Brown, but may have murdered up to as many as twenty women. The alligators were taken to the San Antonio Zoo and Joe Ball became a notorious legend, called the Bluebeard of South Texas. Although the facts and exact number of victims may never be known, one thing is for certain, Joe Ball was a monster who preyed on women.

Conclusion

It's important to note that the serial killers in this book created ripple effects of grief that still resonate today. Our hope is that we illuminated some of their victims' stories while delving into the history and circumstance of when they were alive. Whatever their background, the killers within this book were creators of their own monstrosity, both Dr. Frankenstein and his murderous creature in one. We like it better when the movie lights shine on the make-believe horror, when the good guys win, and when the only upset is not enough butter on our popcorn. There has been enough true horror to last cinematic lifetimes.

If you want to learn more about horror films and the people and stories behind them, read more of our series: *The Science of Monsters*, *The Science of Women in Horror*, and *The Science of Stephen King*.

Acknowledgments

Thank you to Nicole Mele and everyone at Skyhorse!

Thank you to everyone who agreed to be interviewed including Sara, Taneli, Olivia, Mary Kay, Kellie, Lisa B., Janja, Robert, Zach, Rachel, Dave, and Lisa S.

Thank you to Jenny, Nate, Shelley, Nancy, Luke, and Mark for all your help. And as always to our Rewinders, we'll see you in the horror section!

About the Authors

Meg Hafdahl (left) and Kelly Florence

Horror and suspense author Meg Hafdahl is the creator of numerous stories and books. Her fiction has appeared in anthologies such as *Eve's Requiem: Tales of Women, Mystery, and Horror* and *Eclectically Criminal*. Her work has been produced for audio by The Wicked Library and The Lift, and she is the author of three popular short story collections including *Twisted Reveries: Thirteen Tales of the Macabre*. Meg is also the author of the novels *All the Devils, The Darkest Hunger, Daughters of Darkness* and *Her Dark Inheritance*, called "an intricate tale of betrayal, murder, and small-town intrigue" by *Horror Addicts* and "every bit as page turning as any King novel" by *Rochester Women Magazine*. Meg lives in the snowy bluffs of Minnesota.

Kelly Florence teaches communication at Lake Superior College in Duluth, Minnesota, and is the creator of the *Be a Better Communicator* podcast. She received her BA in theater at the University of Minnesota-Duluth and earned her MA in communicating arts at the University of

Wisconsin-Superior. She has written, directed, produced, choreographed, and stage managed for dozens of productions in Minnesota including *Carrie: The Musical* for Rubber Chicken Theatre and *Treasure Island* for Wise Fool Theater. She is passionate about female representation in all media and particularly the horror genre.

Together, Kelly and Meg have written four books: *The Science of Monsters*, *The Science of Women in Horror*, *The Science of Stephen King*, *The Science of Serial Killers*, plus the forthcoming *The Science of Witchcraft*. They co-host the Horror Rewind podcast and write and produce horror projects together.

Endnotes

Chapter One: The Legend of Lizzie Borden (Lizzie Borden)
1. McClure, Rhonda. (April 2004) "Unburying the Hatchet." *Family Tree Magazine*.
2. (2021) "Hydrogen Cyanide." *CDC.gov*.
3. Carlisle, Marcia L. (July 1992) "What Made Lizzie Borden Kill?" *American Heritage Magazine*.
4. Keating, Dan. (May 7, 2015) "The Weapons Men and Women Most Often Use to kill." *The Washington Post*.
5. (2021) "19th Century Mourning." *National Museum of Funeral History*.
6. Hageman, William. (October 9, 2005) "Death in the Living Room." *Chicago Tribune*.
7. (2021) "Victorian Funeral Customs and Superstitions." *Friends of Oak Grove Cemetery.org*.

Chapter Two: From Hell (Jack the Ripper)
1. (2021) "How the Police Investigated the Jack the Ripper Crimes." *JacktheRipper. org*.
2. Thornbury, Walter. (1878) "'Whitechapel', in *Old and New London: Volume 2*." *British History Online*.
3. Sanders, Teela et al. (July 2017) "Reviewing the Occupational Risks of Sex Workers in Comparison to Other 'Risky' Professions." *University of Leicester*.
4. Janos, Adam. (February 12, 2020) "Why Are Sex Workers Often a Serial Killer's Victim of Choice?" *AETV.com*.
5. Janos, Adam. (February 12, 2020) "Why Are Sex Workers Often a Serial Killer's Victim of Choice?" *AETV.com*.
6. Barber, Mike. (February 19, 2003) "Part 4: Serial killers prey on 'the less dead'" *Seattle Post-Intelligencer Reporter*.
7. (2019) Porritt, Lee. "The Curious Case of Henry Grey's Head." *Ladyjanegreyrevisited.com*.
8. (1978) "Victorian Opium Eating: Responses to Opiate Use in Nineteenth-Century England." *Indiana University Press*.
9. (1978) "Victorian Opium Eating: Responses to Opiate Use in Nineteenth-Century England." *Indiana University Press*.
10. Castelow, Ellen. (2016) "Opium in Victorian Britain." *Historic-UK.com*.
11. (2020) "Jack the Ripper Tour." *Jacktherippertour.com*.
12. (2020) "The Ten Bells: A Haunted Hotspot?" *Jacktherippertour.com*.

Chapter Three: The Devil in the White City (H. H. Holmes)
1. (2020) "H.H Holmes." *Famous People.*
2. Pawlak, Debra. (June 11, 2008) "American Gothic: The Strange Life of H. H. Holmes." *Themediadrome.com.*
3. Pawlak, Debra. (June 11, 2008) "American Gothic: The Strange Life of H. H. Holmes." *Themediadrome.com.*
4. Wigington, Patti. (June 30, 2019) "H. H. Holmes: King of the Murder Castle." *Thoughtco.com.*
5. (May 2, 2017) "H. H. Holmes: The Victims of Chicago's First Serial Murderer." *Biography.com.*
6. (2020) "The Bertillon System." Division of Criminal Justice Services.
7. 2020) "The Bertillon System." Division of Criminal Justice Services.
8. (2020) "The Bertillon System." *Crimialjustice.gov.*

Chapter Four: Lake Bodom (Unsolved)
1. (2021) "An Introduction to Crime Scene Reconstruction for the Criminal Profiler." *Crime and Clues.com.*
2. (2020) "Crime in Finland." *Intermin.fi.*
3. (November 9, 2020) "Crime in the United States." *Statista.com.*
4. Ferchland, William. (February 6, 2006) "Lake's Depths Hold Many Dead Bodies." *Tahoe Daily Tribune.*
5. (April 28, 2015) "The Lake Bodom Murders." *Imgur.com.*
6. April 28, 2015) "The Lake Bodom Murders." *Imgur.com.*

Chapter Five: Rillington Place (John Christie)
1. Merryweather, Cherish. (February 11, 2020) "Top 10 Gruesome Ways Serial Killers Disposed of Their Victims." *Listverse.com.*
2. Benwell, Max (24 November 2016) "Rillington Place: What John Christie's Residential Burial Ground Looks Like Now." *The Independent.*
3. Abbott, Kate. (November 29, 2016) "Samantha Morton." *The Guardian.*
4. (2019) "Dismissed." *Today.com.*
5. Bearak J, Popinchalk A, Ganatra B, Moller A-B, Tunçalp Ö, Beavin C, Kwok L, Alkema L.(September 2020) "Unintended Pregnancy and Abortion by Income, Region, and the Legal Status of Abortion." *Lancet Global Health.*
6. Haddad, MD, Lisa B. (2009) "Unsafe Abortion: Unnecessary Maternal Mortality." *Obstetrics and Gynecology.*
7. (2021) "Strangulation." Winchester Hospital.org.
8. (October 11, 2019) "John Christie." *Biography.com.*
9. Palmer, Brian. (December 10, 2009) "Soluble Dilemma." *Slate.com.*
10. (2021) "The Stages of Human Decomposition." *Aftermath.com.*
11. McClelland, PhD., John, Watson, PhD., James T. "Distinguishing Human from Non-Human Animal Bone." *Arizona State Museum.*
12. (2021) "Evidence That May Be Collected." *Forensic Science Simplified.org.*
13. Simpson (Professor), Keith (1978) "Forty Years of Murder." *Grafton Books, London.*

Chapter Six: Wolf Creek (Ivan Milat)

1. (April 14, 2020) "Australia 2020 Crime & Safety Report: Melbourne." *OSAC.gov.*
2. Aiton, Katie Scott. (August 22, 2017) "How to Survive a Road Trip Australia." *Matador Network.*
3. Hand, Eric. (June 26, 2015) "Earth's Colossal Crater Count Complete." *ScienceMag.org.*
4. Nelson, Stephen A. (April 27, 2018) "Meteorites, Impacts, and Mass Extinction." *Tulane University.*
5. (2021) "Meteorite Facts." *The Planets.org.*
6. Freeman, Shanna. (2020) "How Forensic Dentistry Works." *How Stuff Works.com.*
7. (2021) "Stabbing." BrainAndSpinalCord.org.
8. Venman, Kate. (October 28, 2019) "The Full Scope of Ivan Milat's Crimes Will Give You Chills." *GOAT.com.*
9. Clemens, Daniel. (June 21, 2018) "Understanding Link Analysis and Using it in Investigations." *Shadow Dragon.*
10. (2021) "Ivan Milat." *Wikipedia.*
11. (2021) "Shoeprints and Tire Tracks." *Bureau of Criminal Apprehension.*
12. (2021) "Inside the Science of Memory." *John Hopkins University.*
13. (2021) "Murder of Peter Falconio." *Wikipedia.*

Chapter Seven: Monster (Aileen Wuornos)

1. Lyons, Joseph D. (December 3, 2015) "Some Stats on Female Murderers." *Bustle.com.*
2. Blonigen, Daniel M. (July 2003) "A Twin Study of Self-Reported Psychopathic Personality Traits." *Personality and Individual Differences.*
3. (2021)"murderpedia.org/female." CC BY-SA 4.0 *Wikimedia Commons.*
4. Edwards, Jim. (November 24, 2016) "'The Hare Psychopathy Checklist': The test that will tell you if someone is a sociopath." *Business Insider.com.*
5. (April 27, 2017) "Aileen Wuornos." *Biography.com.*
6. Aroncyzk, Melissa. (January 4, 2004) "When Charlize Theron Mined a Deep Well of Darkness to take Grim Lead in Monster." *Toronto Star.*
7. Aroncyzk, Melissa. (January 4, 2004) "When Charlize Theron Mined a Deep Well of Darkness to take Grim Lead in *Monster.*" *Toronto Star.*
8. Stossel, John. (January 5, 2006) "How True is *Monster?*" *ABC News.*
9. (2020) "Lethal Injection." *Death Penalty Info.org.*
10. Segura, Liliana. (February 7, 2019) "Ohio's Governor Stopped an Execution Over Fears it would Feel Like Waterboarding." *The Intercept.*
11. Katz, Josh. (September 10, 2011) "On This Day: France Implements the Final Execution By Guillotine." *Findingdulcenia.com.*

Chapter Eight: Arsenic and Old Lace (Amy Archer Gilligan)

1. Soloway, Rose Ann Gould. (2021) "Black Widow Spiders." *Poison.org.*
2. Szalay, Jessie. (October 30, 2014) "Black Widow Spider Facts." *LiveScience.com.*

3. Houghton, Kristen. (June 22, 2016) "The Real Serial Killer Behind the Play Arsenic and Old Lace." *CriminalElement.com*.
4. Fowler, Dave. (2019) "Oliver Cromwell's Head." *Olivercromwell.net*.
5. (April 4, 2018) "Facts About Strychnine." *CDC.gov*.
6. Gunter, Matthew C. (2012) "The Capra Touch: A Study of the Director's Hollywood Classics and War Documentaries 1934–1945." McFarland and Co Publishers.

Chapter Nine: America's First Female Serial Killer (Jane Toppan)
1. McLeoad, Saul. (2018) "Nature vs. Nurture in Psychology." *Simply Psychology*.
2. (2021) "The Balance of Passions." *U.S. National Library of Medicine*.
3. Davainis, Dava. (December 16, 2019) "The Boston Female Asylum." *Mastatelibrary*.
4. Jenkins, John Phillip. (December 14, 2020) "Harold Shipman." *Britannica*.

Chapter Ten: The Stepfather (John List)
1. Collins, Katie. (August 13, 2013) "Study: Family Killers are Usually Men and Fit One of Four Distinct Profiles." *Wired*.
2. Stewart, Peter. (2003) "Statues in Roman Society: Representation and Response." *Oxford University Press*.
3. Satish, Sandeep. (April 3, 2015) "Autodidactism Or The Unconventional Life." *Soulsophy.com*.
4. Moran, Robert. (July 28th, 2011) "Obituary: Frank Bender, 70, Recomposer of the Dead." *The Philadelphia Inquirer*.
5. Stout, David. (June 20, 2011) "Recomposing Life's Details from Scraps." *The New York Times*.
6. (2021) "17 Fugitives Caught With the Help of America's Most Wanted." *CBS News*.
7. Perry, Douglas. (May 17, 2019). "As New Evidence Upends D.B. Cooper Case, the (Un)usual Suspects Continue to Fuel the Legend." *The Oregonian*.

Chapter Eleven: I'll Be Gone in the Dark (The Golden State Killer)
1. McNamara, Michelle. (2018) *I'll Be Gone in the Dark*. Harper Collins.
2. (August 17, 2020) "Our scars remain: Golden State Killer's victims to tell their stories in court all week." *The Los Angeles Times*.
3. McNamara, Michelle. (February 27, 2013) "The Five Most Popular Myths About the Golden State Killer Case." *Los Angeles Magazine*.
4. Zhang, Sarah. (April 27, 2018) "How a Genealogy Website Led to the Alleged Golden State Killer." *The Atlantic*.
5. Molteni, Megan. (December 26, 2018) "The Future of Crime-Fighting Is Family Tree Forensics." *Wired*.
6. Holes, Paul. (2019) "100 Most Influential People of 2019." *Time*.

Chapter Twelve: My Friend Dahmer (Jeffrey Dahmer)

1. (April 27, 2017) "Jeffrey Dahmer." *Biography.com*.
2. (2020) "10 Little-Known Facts About the Murderous Trio." *True Crime Magazine*.
3. (2021) "Jeffrey Dahmer is Caught." *History.com*.

Chapter Thirteen: Summer of Sam (David Berkowitz)

1. Metcalfe, Nasser. (July 1999) "Spike Lee on SOS." *Black Film.com*.
2. Burke M, Hsiang SM, Miguel E. (2015) "Climate and Conflict." *Ann Rev Econom*.
3. Hancock PA, Vasmatzidiis I. (2003) "Effects of Heat Stress on Cognitive Performance: the Current State of Knowledge." *Int J Hyperthermia*.
4. (March 2017) "US Power Outage Statistics." *Generator Source.com*.
5. (2021) "Cautionary Tale." *Wikipedia*.
6. Zaraska, Marta. (October 11, 2017) "The Sense of Smell in Humans is More Powerful Than We Think." Discover Magazine.
7. Macmillan, Amanda. (November 10, 2017) "Is Sex Addiction Real? Here's What Experts Say." *Time*.
8. LaBianca, Juliana. (April 9, 2020) "Here's What Your Handwriting Says About You." *Reader's Digest.com*.
9. Gibson, Dirk C. (2010). "Clues from Killers: Serial Murder and Crime Scene Messages." *Westport, CT: Praeger*
10. Iulia, Crineanu. (2008) "The Writing of Criminal Minds Criminology and Handwriting Analysis." *International Journal of Criminal Investigation*. Volume 1, Issue 4.
11. Hudson Jr, David L. (March 2012) "'Son of Sam' Laws." *Freedom Forum Institute*.
12. Thorpe, JR. (July 26, 2016) "Why Are People Attracted to Certain Hair Colors?" *Bustle*.
13. Yablon, Alex. (June 21, 2017) "The Simple Physics That Make Some Bullets Deadlier Than Others." *The Trace.org*.
14. Vella MD, Michael A. et al. (November 20, 2019) "Long-term Functional, Psychological, Emotional, and Social Outcomes in Survivors of Firearm Injuries." *JAMA Network*.
15. Bonn PhD, Scott A. (February 10, 2014) "Inside the Mind of Serial Killer 'Son of Sam.'" *Psychology Today*.
16. Sanders, John Vincent. (August 2002) "I am the Son of Sam!" *Fortean Times*.
17. Salleh, Anna. (March 13, 2016) "What Happens in Our Brains When We Hallucinate?" *ABC Science*.

Chapter Fourteen: Dear Mr. Gacy (John Wayne Gacy)

1. Holland, Kimberly. (August 30, 2019) "Understanding Age Regression." *Healthline.com*.
2. (2021) "John Wayne Gacy." *Wikipedia*.
3. Grusec PhD, Joan E. Danyliuk, Tanya. (December 2014) "Parents' Attitudes and Beliefs: Their Impact on Children's Development." *University of Toronto, Canada*.

4. Sullivan, Terry; Maiken, Peter T. (2000) "Killer Clown: The John Wayne Gacy Murders." *Pinnacle.*

5. (2021) "Antisocial Personality Disorder." *Mayo Clinic.org.*

6. Gluck, S. (December 4, 2014) "Famous People with Antisocial Personality Disorder." *Healthy Place.*

7. (2021) "John Wayne Gacy." *Wikipedia.*

8. Tedeschi, Christopher M. (August 16, 2018) "Life After Drowning." *Scientific American.*

9. Lyle, DP. (2015) "31: Body Disposal." *DP Lyle MD.com.*

10. (2021) "Quicklime." *SMA Mineral.*

11. (2021) "Alkaline Hydrolysis." *Wikipedia.*

12. Polkes, Aliza. (March 20, 2017) "9 Little-Known Facts About John Wayne Gacy." *The Lineup.*

Chapter Fifteen: A Good Marriage (The BTK Killer)

1. King, Stephen. (2010) "A Good Marriage: Inspiration." *Stephen King.com.*

2. Sherman, Beth. (November 29, 1992) "Leading a Double Life." *Los Angeles Times.com*

3. (February 14, 2020) "Dennis Rader." *Biography.com.*

4. Wenzl, Roy. (September 25, 2014) "BTK's Daughter: Stephen King 'Exploiting My Father's 10 Victims and Their Families' With Movie." *The Wichita Eagle.*

5. Douglas, John. (October 6, 2012) "Why Killers Take Trophies." *Mindhunters Inc.com.*

6. (May 7, 2007) "Serial Killer Souvenirs." *Sword and Scale.com.*

7. Abraham, Micah. (November 25, 2020) "How Anxiety Can Create Hallucinations." *Calm Clinic.com.*

8. Goleman, Daniel (August 7, 1991) "Child's Love of Cruelty May Hint at the Future Killer". *New York Times.*

9. Reeve, Elspeth. (June 12, 2013) "Why Your Metadata Is Your Every Move." *The Atlantic.*

10. (July 28, 2020) "How Forensic Specialists Solve Crime Through Metadata." *Caseguard.com.*

11. Lees, Jonathan. (October 3, 2014) "Stephen King, Marriage Counselor? Yes... If Your Lover's a Psycho Killer." *Complex.com.*

Chapter Sixteen: Scream (The Gainesville Ripper)

1. Leusner, Jim. 9 (March 11, 1994) "On Tape, Rolling Pours Out His Tale of Torment." *Orlando Sentinel.*

2. Bonn, Scott A. (June 29, 2015) "Serial Killers: Modus Operandi, Signature, Staging & Posing: Understanding and classifying serial killer crime scenes." *Psychology Today.*

3. Hunter, George. (June 11, 2019) "Serial killer experts say kneeling corpses are symbolic." *The Detroit News.*

4. Griffiths, Mark D. (October 18, 2013) "Passion Victim." *Psychology Today.*

5. Crabbe, Nathan. (August 27, 2005) "Waiting For Justice: Crime Pays." *Gainesville Sun.*

6. Hintz, Charlie. (2017) "The Head of German Serial Killer Peter Kurten in Wisconsin Dells." *Cultofweird.com.*

7. Bonn, Scott A. (June 29, 2015) "Serial Killers: Modus Operandi, Signature, Staging & Posing: Understanding and classifying serial killer crime scenes." *Psychology Today.*

8. La Ganga, Maria L. (August 10, 2019) "At the Museum of Death, every day is like a Charles Manson anniversary." *Los Angeles Times.*

Chapter Seventeen: Rope (Leopold and Loeb)

1. Ramsland, Katherine. (May 29, 2016) "To Be or Not to Be: Philosophical Serial Killers." *Psychology Today.*

2. Magnus, Bernd. (October 11, 2020) "Friedrich Nietzsche." *Britannica.com.*

3. Cybulska, Eva. (2012) "Nietzsche's Ubermensch: A Hero of Our Time?" *Philosophy Now.*

4. Cybulska, Eva. (2012) "Nietzsche's Ubermensch: A Hero of Our Time?" *Philosophy Now.*

5. Hindley, Meredith. (2012) "Nietzsche is Dead." *Humanities.*

6. Walsh, Michael. (October 3, 2014) "Dexter Obsessed Teen Jailed for Stabbing, Dismembering Girlfriend." *New York Daily News.*

Chapter Eighteen: Extremely Wicked, Shockingly Evil and Vile (Ted Bundy)

1. Nolan, Amy. (2019) "People have some STRONG opinions about the judge in Ted Bundy's trial after watching the docu-series." *Her.*

2. Arntfield, Michael and Danesi, Marcel. (2017) "Murder in Plain English: From Manifestos to Memes—Looking at Murder through the Words of Killers." *Prometheus.*

3. Berney, Barbara. (1993) "Round and Round it Goes: The Epidemiology of Childhood Lead Poisoning 1950 to 1990." *The Milbank Quarterly, Vol 71. No 1.*

4. Dignam PhD, Timothy. (2019) "Control of Lead Sources in the United States, 1970–2017: Public Health Progress and Current Challenges to Eliminating Lead Exposure." *Public Health Manag Pract.*

5. Drum, Kevin. (February 1, 2018) "An Updated Lead Crime Update for 2018." *Mother Jones.*

6. Arnold, Carrie. (May 31, 2017) "The Man Who Warned the World About Lead." *PBS.Org.*

7. Carey, Benedict. (July 27, 2017) "Dr. Herbert Needleman, Who Saw Lead's Wider Harm to Children, Dies at 89." *The New York Times.*

8. Drum, Kevin. (February 1, 2018) "An Updated Lead-Crime Roundup for 2018." *Mother Jones.*

Chapter Nineteen: American Crime Story: The Assassination of Gianni Versace (Andrew Cunanan)

1. Swinson, Brock. (September 13, 2018) "Unmasking Versus Sensation." *Creative Screenwriting.com.*
2. Vargas, Chanel (February 28, 2018) "Who is Andrew Cunanan, the Man who Murdered Gianni Versace?" *Town and Country.*
3. Stillman, Jessica. (October 2, 2018) "The Science of Lying." *Inc.com.*
4. (2021) "Top 10 Signs That Someone Is Lying." *ForensicColleges.com.*
5. Snitzer, Zachary. ((2021) "The Effects of Painkillers on the Brain and Body." *Maryland Addiction Recovery.com.*
6. Thomas MD, Scot. (September 3, 2019) "Dangers of Mixing Alcohol and Opiates, including Hydrocodone, Oxycodone and Morphine." *American Addiction Centers.org.*
7. (2020) "Can Dogs Sense Fighting?" *Wag Walking.com.*
8. Spencer, Bill. (May 4, 2017) "Would Your Dog Attack An Intruder?" *Houston.com.*
9. (May 27, 2020) "Dog Bite Statistics." *Canine Journal.com.*
10. (2021) "Domestic Violence Statistics." *The Hotline.org.*
11. Roush, Melanie. (2012) "Different Mindset: Negotiation Challenges for Today's Critical Incident Responders." *Gazette, Volume:64.*
12. Taylor, Andrew. (2021) "Surviving a Hostage Situation." *IHL.State.ms.us.*

Chapter Twenty: Henry: Portrait of a Serial Killer (Henry Lee Lucas)

1. Knight, Jacob. (October 17, 2016) "Michael Rooker Talks HENRY: PORTRAIT OF A SERIAL KILLER On Its 30th Anniversary." *Birth. Movies. Death.com.*
2. Chris Jozefowicz. (October 1, 2003) "Understanding Compulsive Liars." *Psychology Today.*
3. (2021) "Henry Lee Lucas." *Wikipedia.*
4. (2021) "False Confessions & Recording of Custodial Interrogations." *The Innocence Project.*
5. Wade, Bethany. (April 10, 2020) "Portrait of a Serial Killer: Henry Lee Lucas and His Bloodsoaked Legacy." *Film Daily.*
6. Pasteur, Louis. (1854) "Lecture."
7. Gajanan, Mahita. (December 6, 2019) "The Story of Henry Lee Lucas, the Notorious Subject of Netflix's The Confession Killer." *Time.*
8. Mason, Steven. (January 11, 2018) "Investigative Timelines in Criminal Defense Investigations." *Pursuit Magazine.*
9. Black, Lisa. (July 11, 2010) "What Causes People to Give False Confessions?" *Chicago Tribune.*
10. (2021) "Ottis Toole." *Wikipedia.*
11. Robinson, Abby. (January 22, 2019) "7 Actors Who Took 'The Method' Waaay Too Far." *Digital Spy.*
12. Cole, Megan. (November 28, 2017) "Scientific Method Acting." *UCI News.*

Chapter Twenty-One: American Predator (Israel Keyes)
1. (May 21, 2020) "Israel Keyes." *Biography.com.*
2. (2020) "What the Research Says on Socialization." *Responsible Homeschooling. org.*
3. Niedospial, Laurel. (May 12, 2018) "What Homeschooling Gets Wrong About Socialization." *Popsugar.com.*
4. Diament, Michelle. (November 21, 2008) "Friends Make You Smart." *AARP Bulletin.*
5. Heiphetz, Larisa et al. (May 1, 2013) "The Development of Reasoning about Beliefs: Fact, Preference, and Ideology." *J Exp Soc Psychol.*
6. Fields, PhD. R. Douglas. (November 12, 2018) "Inside the Mind of an Arsonist." *Psychology Today.*
7. (2020) "Arson Information." *Illinois.gov.*
8. Ferron, Emily. (October 12, 2020) "2019 Safety.com Home Security Report." *Safety.com.*
9. (August 13, 2013) "Seeking the Public's Assistance." *FBI.gov.*
10. Callahan, Maureen. (2019) "American Predator." *Viking Press.*
11. (March 8, 2009) "Humans Can Sense 'Smell Of Fear' In Sweat, Psychologist Says." *Rice University.*
12. (2021) "Kleptomania." *Mayo Clinic.org.*
13. D'Oro, Rachel, Cohen, Sharon. (March 24, 2019) "Trying to Unlock Secrets of Dead Serial Killer Israel Keyes." *Mass Live.com.*

Chapter Twenty-Two: Zodiac (The Zodiac Killer)
1. Mackintosh, Prudence. (December 2014) "Texarkana Murder Mystery." *Texas Monthly.*
2. (April 20, 1970) "Letter sent to San Francisco Chronicle."
3. (July 31, 1969) "Letter sent to San Francisco Chronicle."
4. (2021) "Homophonic Cipher." *Dcode.fr.*
5. Suetonius. (121 AD) "Life of Julius Caesar."
6. Singh, Simon. (2009) The Code Book: The Science of Secrecy from Ancient Egypt to Quantum Cryptography."
7. (August 22, 2018) Bauer, Craig P. "The Zodiac Ciphers: What We Know" *History.com."*
8. (August 8, 1969) "Zodiac 408 Cipher."
9. (December 2020) "340 Cipher."
10. Asemlasah, Leah and Mossburg, Cheri. (December 11, 2020) "After 51 years, the Zodiac Killer's cipher has been solved by amateur code breakers." *CNN.*
11. Pelling, Nick. (March 31, 2013) "The Unabomber Cipher Journal." *Cipher Mysteries.com.*

Chapter Twenty-Three: Wind River (The Highway of Tears)
1. La Jeunesse, Marilyn. (August 29, 2019) "Unsolved Crimes by Serial Killers Who Might Never be Caught." *Insider.*

2. Sabo, Don. (January 18, 2019) "Highway of Tears." *The Canadian Encyclopedia.*
3. (May 2019) "Indigenous Overrepresentation in the Criminal Justice System." *Canada Department of Justice.*
4. Clay, Allison. (December 8, 2015) "The Highway of Tears: A Case Study." *Missing Indegenous Women.com.*
5. Bonn, Scott PhD. (December 7, 2018) "What Is The Serial Killer Capital of the U.S.?" *Psychology Today.*
6. Watts, Sarah. (May 14, 2018) "Why Are There More Serial Killers in the U.S. Than Any Other Country?" *A&ETV.com.*
7. (November 12, 2019) "Which State Has Produced the Most Serial Killers?" *Crimecapsule.com.*

Chapter Twenty-Four: The Man from the Train (Unsolved)
1. (2021) Scranton, Laird. "Shango." *Britanica.com.*
2. (2021) "Labrys Symbol, Its Meaning, History and Origins." *Mythologian.net.*
3. (2014) Meij, Harold. "The Symbolism of the Axe." *The Masonic Trowel.com.*
4. (March 13, 1919) "The Axeman Letter." *The Times Picaune.*
5. (May 26, 2020) "The Axeman of New Orleans." *NOLAGhosts.com.*
6. (2021) "BTK and Do Serial Killers Want to Get Caught?" *Crime Investigation.co.uk.*
7. (July 8, 1995) Brazil, Jeff. "Killers to the Police: Catch Me if You Can: From Jack the Ripper to the Unabomber, some criminals like to taunt their pursuers." *LA Times.*
8. (2017) James, Bill and McCarthy James, Rachel. *The Man from the Train: The Solving of a Century Old Serial Killer Mystery.* 2017. Simon and Schuster.

Chapter Twenty-Five: The Strangers (Charles Manson)
1. Catalano PhD., Shannan. (September 2010) "Victimization During Household Burglary." BJS.gov.
2. Turek, Ryan. (May 26, 2008) "The Strangers' Bryan Bertino (Pt. 2)." *Comingsoon. net.*
3. (2018) "The National 911 Program Celebrates 50 Years of 911." *911.gov.*
4. Donaghey, River. (2020) "How Charles Manson Put an End to the Hippie Movement." *Vice.*
5. (2012) "The Strangers Production Notes." *Cinema Review.*
6. Ramirez, Ainissa. (October 17, 2014) "The Science of Fear." *Edutopia.*
7. (2020) "Citizens Band Radio." *Walcott Radio.com.*
8. (2021) "Keddie Murders." *Wikipedia.*

Chapter Twenty-Six: Speck (Richard Speck)
1. (2021) "Chicago Strangler." *Wikipedia.*
2. (July 14, 2016) "Slain Nurses Remembered on 50th Anniversary of Richard Speck Murders." *ABC 7, Chicago.com.*
3. (2020) "Heroin Addiction." Advanced Recovery Systems.com.

4. Allodi FA.(June 1994) "Post-traumatic Stress Disorder in Hostages and Victims of Torture." *Psychiatric Clinics of North America.*
5. Conger, Cristen. (2011) "How Police Sketches Work." *How Stuff Works.com.*
6. Conger, Cristen. (2011) "How Police Sketches Work." *How Stuff Works.com.*
7. Bareket, O., Kahalon, R., Shnabel, N., & Glick, P. (2018) "The Madonna-Whore Dichotomy: Men who perceive women's nurturance and sexuality as mutually exclusive endorse patriarchy and show lower relationship satisfaction.*" Sex Roles: A Journal of Research.*
8. Lenroot, MD, Rhoshel K. (2012) "XYY Syndrome.*" National Organization for Rare Disorders.*
9. Breo, Daniel L.; Martin, William J.; Kunkle, Bill (1993). "The Crime of the Century: Richard Speck and the Murders That Shocked a Nation." New York City: *Bantam Books.*
10. (January 31, 2015) "Organic Mental Disorder." *White Swan Foundation.org.*
11. (2021) "Speck: Trivia.*" IMDB.com.*
12. Pyszora, Natalie M. et al. (June 2014) "Amnesia for Violent Offenses: Factors Underlying Memory Loss and Recovery.*" Journal of the American Academy of Psychiatry and the Law Online.*
13. (2021) "Richard Speck." *Wikipedia.*

Chapter Twenty-Seven: American Horror Story: 1984 (The Night Stalker)

1. Simon, Amie. (June 6, 2016) "Imaginary SIFF Interview: 5 questions with Lou Diamond Phillips {+ a bonus question with Megan Griffiths!} re: The Night Stalker." *Three Imaginary Girls.com.*
2. Stone, Michael H. (May 3, 2007)) "Personality-Disordered Patients: Treatable and Untreatable." *American Psychiatric Pub.*
3. Wilson, PhD. Eric G. (November 8, 2011) "The Moral of the Morbid." *Psychology Today.*
4. (2020) "For Caregivers." Child Witness to Violence Project.org.
5. Carlo, Philip (1996)."The Night Stalker: The Life and Crimes of Richard Ramirez." New York, New York: *Kensington Publishing Corp.*
6. (January 10, 2015) "Martin Place Siege Victim Katrina Dawson Struck by a Police Bullet, Investigations Show." *Sydney Morning Herald.*
7. Nirenber, Michael. (January 2016) "Gait, Footprints, and Footwear: How Forensic Podiatry Can Identify Criminals." *Police Chief Magazine.*

Chapter Twenty-Eight: The Amityville Horror (Ronald DeFeo Jr.)

1. (February 27, 2012) "New Evidence Raises Questions In Decades-Old Amityville Horror Murders." *CBSN, New York.*
2. Naxalovka123, CC BY-SA 4.0, via Wikimedia Commons.
3. (December 2020) "Detecting Air Leaks." *Energy.gov.*
4. Swedenborg, Emanuel. (1846) "The Spiritual Diary of Emanuel Swedenborg." *The Swedenborg Society, London.*

5. Andrews, Jeff. (October 25, 2018) "Buying the Nightmare on Elm Street." *Curbed.com*.
6. Steele, Chandra. "October 31, 2014) "7 Surprising Ghost-Hunting Gadgets." *PC Magazine*.
7. Smallegange, Renate C., Niels O. Verhulst, and Willem Takken.(2011) "Sweaty skin: an invitation to bite?." *Trends in Parasitology 27.4*.
8. Hartung, Edward Frederick. (1935). "St. Francis and Medieval Medicine." *Annals of Medical History 7*: 85–91.
9. Smith-Randolph, Walter. (September 27, 2019) "Exorcist Explains Why Demand For Exorcisms Has Risen." *ABC News*.
10. Booth, Jessica. (June 26, 2019) "5 Spooky Things Your Dog Can Sense That You Can't." *Bustle*.
11. (2017) "Can Dogs Sense the Supernatural?" *Animal Planet.com*.
12. Heimbuch, Jaymi. October 14, 2020) "6 Medical Conditions That Dogs Can Sniff Out." *Treehugger.com*.
13. Darling, Ruthie. (November 1, 2020) "This is Why You Often Wake Up Around 3 am." *Considerable.com*.
14. Dyer, Dr. Wayne. (October 26, 2012) *Dr. Wayne Dyer Facebook Post*.
15. (July 17, 2020) "What Is Carbon Monoxide?" *CDC.gov*.
16. (November 22, 2020) "Handling Steering Failure: Power Steering, Common Causes & What To Do." *ePermitTest.com*.
17. Feuerstein, MD, Seth. (September 2005) "The Insanity Defense." *Psychiatry*.
18. (2020) "Paranoia." *WebMD*.

Chapter Twenty-Nine: The Boston Strangler (Albert DeSalvo)

1. Ebert, Roger. (October 22, 1968) "The Boston Strangler." *Chicago Sun Times*.
2. Campbell, Duncan. (September 1, 2015) "Why Involve Psychics in Police Investigations When Their Track Record is So Poor?" *The Guardian*.
3. (October 30, 2019) "Using Psychic Abilities to Solve Potential Murders." *Ripley's.com*.
4. Schwartzbach, Micah. (2021) "Is a Confession Alone Enough to Convict a Defendant?" *NOLO.com*.
5. Johnson, Scott Allen. (2011) "Power Reassurance Rapist." *Forensic Consultation.org*.
6. (2021) RAINN.org.
7. Bidgood, Jess. (July 11, 2013) "50 Years Later, a Break in a Boston Strangler Case." *The New York Times*.
8. Frank, Gerold (1966) "The Boston Strangler." *New York City: Signet*.

Chapter Thirty: Eaten Alive (Joe Ball)

1. Redhead, Harry. (May 24, 2015) "Alligator Takes on Truck…and Wins." *Metro*.
2. Englund, Robert. (2012) "Eaten Alive DVD Extras." *YouTube*.
3. (2021) "Bootlegging." *Britannica.com*.

4. (2020) "What Is the Corpus Delicti Rule About Sufficient Evidence?" *Greg Hills and Associates.com.*
5. Ruppenthal, Alex. (July 11, 2019) "Why Alligators Don't Make Good Pets."
6. Hoover, Marc. (August 22, 2019) "Joseph Briggen and His Hogs he Used to Dispose of Dead Bodies." *The Clermont Sun.*
7. (September 15, 2011) "Supersize Crocs." *Nature.*
8. Sczesny, Matt. (December 13, 2019) "Martin County Deputies Warning Pet Owners to be Careful After Deadly Gator Attacks." *WPTV.*
9. Hogenboom, Melissa. (July 21, 2015) "Why Do Crocodiles Attack Humans?" *BBC Earth.*
10. (2020) "Keeping Crocodiles as Pets." *NT.gov.au.*
11. Falcon, Gabe. (January 15, 2007) "Raw Data: Kidnapping Statistics." *CNN.*
12. Ramsland Ph.D., Katherine. (July 31, 2019) "Serial Killer Diaries." *Psychology Today.*
13. Brown, Ford Maddox. (May 20, 2017) "Eaten Alive." *Starburst Magazine.*
14. Murrell M.D., Daniel. (July 17, 2018) "What Happens When You Get an Adrenaline Rush." *Medical News Today.*
15. Wise, Jeff. (December 28, 2009) "When Fear Makes Us Superhuman." *Scientific American.*
16. George, Jane. (February 17, 2006) "Polar Bear No Match for Fearsome Mother in Ivujivik." *Nunatsiaq News.*

Index

Note: Illustrations are indicated by page numbers in *italics*.